AF594939

Low-Cost Inventions and Patents

Low-Cost Inventions and Patents

Editors

Francisco Manzano Agugliaro
Esther Salmerón-Manzano

MDPI • Basel • Beijing • Wuhan • Barcelona • Belgrade • Manchester • Tokyo • Cluj • Tianjin

Editors

Francisco Manzano Agugliaro
Engineering
University of Almeria
Almeria
Spain

Esther Salmerón-Manzano
Faculty of Law
Universidad Internacional de la
Rioja (UNIR)
La Rioja
Spain

Editorial Office
MDPI
St. Alban-Anlage 66
4052 Basel, Switzerland

This is a reprint of articles from the Special Issue published online in the open access journal *Inventions* (ISSN 2411-5134) (available at: www.mdpi.com/journal/inventions/special_issues/inventions_patents).

For citation purposes, cite each article independently as indicated on the article page online and as indicated below:

LastName, A.A.; LastName, B.B.; LastName, C.C. Article Title. *Journal Name* **Year**, *Volume Number*, Page Range.

ISBN 978-3-0365-2989-9 (Hbk)
ISBN 978-3-0365-2988-2 (PDF)

Contents

About the Editors vii

Preface to "Low-Cost Inventions and Patents" ix

Esther Salmerón-Manzano and Francisco Manzano-Agugliaro
Low-Cost Inventions and Patents
Reprinted from: *Inventions* **2022**, *7*, 13, doi:10.3390/inventions7010013 1

George Voudiotis, Sotirios Kontogiannis and Christos Pikridas
Proposed Smart Monitoring System for the Detection of Bee Swarming
Reprinted from: *Inventions* **2021**, *6*, 87, doi:10.3390/inventions6040087 5

Dora Cama-Pinto, Juan Antonio Holgado-Terriza, Miguel Damas-Hermoso, Francisco Gómez-Mula and Alejandro Cama-Pinto
Radio Wave Attenuation Measurement System Based on RSSI for Precision Agriculture: Application to Tomato Greenhouses
Reprinted from: *Inventions* **2021**, *6*, 66, doi:10.3390/inventions6040066 25

Juan D. Borrero
Expanding the Level of Technological Readiness for a Low-Cost Vertical Hydroponic System
Reprinted from: *Inventions* **2021**, *6*, 68, doi:10.3390/inventions6040068 39

Chun Quan Kang, Poh Kiat Ng and Kia Wai Liew
The Conceptual Synthesis and Development of a Multifunctional Lawnmower
Reprinted from: *Inventions* **2021**, *6*, 38, doi:10.3390/inventions6020038 53

Renan Rocha Ribeiro, Elton Bauer and Rodrigo Lameiras
HIGROTERM: An Open-Source and Low-Cost Temperature and Humidity Monitoring System for Laboratory Applications
Reprinted from: *Inventions* **2021**, *6*, 84, doi:10.3390/inventions6040084 79

Zain-Aldeen S. A. Rahman, Basil H. Jasim, Yasir I. A. Al-Yasir, Raed A. Abd-Alhameed and Bilal Naji Alhasnawi
A New No Equilibrium Fractional Order Chaotic System, Dynamical Investigation, Synchronization, and Its Digital Implementation
Reprinted from: *Inventions* **2021**, *6*, 49, doi:10.3390/inventions6030049 97

Daniel Fontoura Barroso, Niklas Epple and Ernst Niederleithinger
A Portable Low-Cost Ultrasound Measurement Device for Concrete Monitoring
Reprinted from: *Inventions* **2021**, *6*, 36, doi:10.3390/inventions6020036 113

Ephraim Bonah Agyekum, Seepana PraveenKumar, Aleksei Eliseev and Vladimir Ivanovich Velkin
Design and Construction of a Novel Simple and Low-Cost Test Bench Point-Absorber Wave Energy Converter Emulator System
Reprinted from: *Inventions* **2021**, *6*, 20, doi:10.3390/inventions6010020 131

About the Editors

Francisco Manzano Agugliaro

Francisco Manzano-Agugliaro, full professor at the Engineering Department in the University of Almeria (Spain), received his M.Sc. in Agricultural Engineering and completed his Ph.D. in Geomatics at the University of Cordoba (Spain). He has published over 180 papers in *JCR* journals, with an H-index of 7. His main interests are energy, sustainability, agronomy, water, and engineering. He has been the supervisor of 31 Ph.D. theses. He has also been the Vice Dean of the Engineering Faculty (2001–2004); the Director of Central Research Services (2016–2019); Ph.D. Program Coordinator for Environmental Engineering (2000 to 2012), Greenhouse Technology, Industrial and Environmental

Engineering (from 2010 to 2021); and General Manager of Infrastructures (from 2019) at University of Almeria. He has won the following awards: Top Reviewer in Cross-Field—September 2019 (Web of Science), 2019 Outstanding Reviewer Award (Energies), 2019 Winner of Sustainability Best Paper Awards.

Esther Salmerón-Manzano

Esther Salmeron-Manzano is a lecturer at the Faculty of Law in the Universidad Internacional de la Rioja (Spain) and Lecturer at Law Department of University of Almeria (Spain). She received her M.Sc. degree in Law and completed her Ph.D. in Law at the University of Almeria (Spain). She has published over 20 papers in *JCR* journals, with an H-index of 11. Her main interests are laws and emerging technologies, and contract law. She has been the supervisor of 52 bachelor's and master's Final Reports. She is an Academic Director of the master's degree in legal consultancy for companies and the master's degree in family law at Universidad Internacional de la Rioja (UNIR).

Preface to "Low-Cost Inventions and Patents"

Today, there are marketable solutions for almost everything, but many of them are not available to everyone due to their cost. We investiagate low-cost solutions to accessible technology resources that one can build by oneself with the right knowledge, based on free technologies and open design, including Opensource. Thus, low-cost inventions are always created in response to a specific need, and even if a thorough analysis of the market situation has not been performed, they are usually at least 10 times cheaper than standard commercial solutions. On the other hand, inventions that give rise to patents can rarely be published, since doing so would invalidate the process of obtaining the patent. For this reason, explaining the development of the invention once it has been patented is suggested.

This book presents examples of the development of low-cost inventions in two fields of engineering: agronomy and monitoring.

Francisco Manzano Agugliaro, Esther Salmerón-Manzano
Editors

MDPI

Editorial

Low-Cost Inventions and Patents

Esther Salmerón-Manzano [1] and Francisco Manzano-Agugliaro [2,*]

1 Faculty of Law, Universidad Internacional de La Rioja (UNIR), Av. de la Paz, 137, 26006 Logroño, Spain; esther.salmeron@unir.net
2 Department of Engineering, University of Almeria, ceiA3, 04120 Almeria, Spain
* Correspondence: fmanzano@ual.es; Tel.: +34-950-015-346

1. Introduction

Inventions have been the technological advances of mankind. There are inventions of all kinds, some of which have lasted hundreds of years or even longer, preserving their essence. For example, the pipe as a means of transporting water, concrete to attach stones in construction, or the pulley to lift heavy objects; all of these are attributed to Roman times, and are still relevant today. Low-cost technologies are expected to be easy to build, have little or no energy consumption, and be easy to maintain and operate. The use of sustainable technologies is essential in order to move towards greater global coverage of technology, and therefore to improve human quality of life. Low-cost products always respond to a specific need, even if no in-depth analysis of the situation or possible solutions have been carried out. It is a consensus in all industrialized countries that patents have a decisive influence on the organization of the economy, as they are a key element in promoting technological innovation. Patents must aim to promote the technological development of countries, starting from their industrial situation.

Usually, for the proposal of a patent, a review of the state of technology on that particular issue must be made. This type of study is not often published in scientific journals, despite the technological value they have. This is why this Special Issue aims to include research works reviewing the state of the art in low-cost technologies used for patent applications. One of the ways that allows the survival of a company, in a world as competitive as the current one, is the promotion of technological innovation through its inventions; this makes it possible to add value to the services and products it can offer. The management of technological innovation requires the implementation of a set of mechanisms, including the protection of intellectual property. Thus, patents, licenses, copyrights, trademarks and trade secrets are some of the legal tools through which inventors and innovators enforce their legitimate right to keep ownership of their inventions.

2. Publications Statistics

This Special Issue has eight published manuscripts. The submitted manuscripts come from nine countries and are summarized in Table 1. For this statistic, only the first affiliation of the authors has been considered. Note that it is common for a manuscript to be signed by more than one author and for authors to belong to different affiliations. The average number of authors per published manuscript in this Special Issue was three authors.

Citation: Salmerón-Manzano, E.; Manzano-Agugliaro, F. Low-Cost Inventions and Patents. *Inventions* **2022**, *7*, 13. https://doi.org/10.3390/inventions7010013

Received: 4 January 2022
Accepted: 10 January 2022
Published: 11 January 2022

Publisher's Note: MDPI stays neutral with regard to jurisdictional claims in published maps and institutional affiliations.

Table 1. Authors' countries: statistics.

Country	Authors
Brazil	3
Colombia	1
Germany	3
Greece	3
Iraq	3
Malaysia	3
Russia	4
Spain	5
UK	2
Total	27

3. Authors' Affiliations

There are 14 different author affiliations for the manuscripts of this Special Issue. Note that only the first affiliation of each author has been considered. Table 2 summarizes the authors and their first affiliations.

Table 2. Authors' affiliations.

Author	First Affiliation	References
Ephraim Bonah Agyekum	Ural Federal University	[1]
Seepana PraveenKumar	Ural Federal University	[1]
Aleksei Eliseev	Ocean Rus Energy	[1]
Vladimir Ivanovich Velkin	Ural Federal University	[1]
Daniel Fontoura Barroso	Bundesanstalt für Materialforschung und -prüfung (BAM)	[2]
Niklas Epple	Bundesanstalt für Materialforschung und -prüfung (BAM)	[2]
Ernst Niederleithinger	Bundesanstalt für Materialforschung und -prüfung (BAM)	[2]
Chun Quan Kang	Multimedia University	[3]
Poh Kiat Ng	Multimedia University	[3]
Kia Wai Liew	Multimedia University	[3]
Zain-Aldeen S. A. Rahman	Southern Technical University	[4]
Basil H. Jasim	University of Basrah	[4]
Yasir I. A. Al-Yasir	University of Bradford	[4]
Raed Abd-Alhameed	University of Bradford	[4]
Bilal Naji Alhasnawi	University of Basrah	[4]
Juan D. Borrero	Huelva University	[5]
Renan Rocha Ribeiro	University of Brasília	[6]
Rodrigo Lameiras	University of Brasília	[6]
Dora Cama-Pinto	University of Granada	[7]
Juan Antonio Holgado-Terriza	University of Granada	[7]
Miguel Damas-Hermoso	University of Granada	[7]
Francisco Gómez-Mula	University of Granada	[7]
Alejandro Cama-Pinto	Universidad de la Costa	[7]
George Voudiotis	University of Ioannina	[8]
Sotirios Kontogiannis	University of Ioannina	[8]
Christos Pikridas	Aristotle University of Thessaloniki	[8]

4. Topics and Keywords

Table 3 summarizes the research conducted by the authors in this Special Issue, by identifying the areas to which they report. It was noted that they have been grouped into two main lines of research: Agronomy and Monitoring. Table 4 summarizes the keywords of the published manuscripts. The most frequent keywords were: IoT [2,7,8] and low-cost [2,6], see Table 4.

Table 3. Topics for low-cost inventions and patents.

Topics	Number of Manuscripts	References
Agronomy	3	[3,5,7]
Monitoring	1	[1,2,4,6–8]

Table 4. Keywords for low-cost inventions and patents.

Keywords	References
wave energy converter; emulator; point absorber; power take-off; hydrodynamics	[1]
low-cost; ultrasound; IoT; non-destructive testing; coda wave interferometry	[2]
conceptual synthesis; design; lawnmower; multifunctionality; usability	[3]
fractional order; dynamics; chaotic; system; synchronization; arduino due	[4]
vertical farming; indoor farming; low-cost vertical farming; small farming; vertical hydroponics; technology readiness level; utility model	[5]
temperature; humidity; low-cost; open-source; Arduino; data-logging; material testing; laboratory; electronics	[6]
wireless sensor networks; WSN; received signal strength indicator; RSSI; Internet of Things; IoT; free space pathloss; smart farming	[7]
beehive-monitoring systems; IoT; convolutional neural networks; image processing; performance evaluation; distributed systems	[8]

Author Contributions: The authors all made equal contributions to this article. All authors have read and agreed to the published version of the manuscript.

Funding: This research received no external funding.

Institutional Review Board Statement: Not applicable.

Informed Consent Statement: Not applicable.

Data Availability Statement: Not applicable.

Acknowledgments: The authors would like to thank to the CIAIMBITAL (University of Almeria, CeiA3) for its support.

Conflicts of Interest: The authors declare no conflict of interest.

References

1. Agyekum, E.B.; PraveenKumar, S.; Eliseev, A.; Velkin, V.I. Design and Construction of a Novel Simple and Low-Cost Test Bench Point-Absorber Wave Energy Converter Emulator System. *Inventions* **2021**, *6*, 20. [CrossRef]
2. Fontoura Barroso, D.; Epple, N.; Niederleithinger, E. A Portable Low-Cost Ultrasound Measurement Device for Concrete Monitoring. *Inventions* **2021**, *6*, 36. [CrossRef]
3. Kang, C.Q.; Ng, P.K.; Liew, K.W. The Conceptual Synthesis and Development of a Multifunctional Lawnmower. *Inventions* **2021**, *6*, 38. [CrossRef]
4. Rahman, Z.-A.S.A.; Jasim, B.H.; Al-Yasir, Y.I.A.; Abd-Alhameed, R.A.; Alhasnawi, B.N. A New No Equilibrium Fractional Order Chaotic System, Dynamical Investigation, Synchronization, and Its Digital Implementation. *Inventions* **2021**, *6*, 49. [CrossRef]
5. Borrero, J.D. Expanding the Level of Technological Readiness for a Low-Cost Vertical Hydroponic System. *Inventions* **2021**, *6*, 68. [CrossRef]
6. Rocha Ribeiro, R.; Bauer, E.; Lameiras, R. HIGROTERM: An Open-Source and Low-Cost Temperature and Humidity Monitoring System for Laboratory Applications. *Inventions* **2021**, *6*, 84. [CrossRef]
7. Cama-Pinto, D.; Holgado-Terriza, J.A.; Damas-Hermoso, M.; Gómez-Mula, F.; Cama-Pinto, A. Radio Wave Attenuation Measurement System Based on RSSI for Precision Agriculture: Application to Tomato Greenhouses. *Inventions* **2021**, *6*, 66. [CrossRef]
8. Voudiotis, G.; Kontogiannis, S.; Pikridas, C. Proposed Smart Monitoring System for the Detection of Bee Swarming. *Inventions* **2021**, *6*, 87. [CrossRef]

Article

Proposed Smart Monitoring System for the Detection of Bee Swarming

George Voudiotis [1], Sotirios Kontogiannis [1,*] and Christos Pikridas [2]

[1] Laboratory Team of Distributed Microcomputer Systems, Department of Mathematics, University of Ioannina, 45110 Ioannina, Greece; georgosvoud@gmail.com

[2] School of Rural and Surveying Engineering, Aristotle University of Thessaloniki, 54124 Thessaloniki, Greece; cpik@topo.auth.gr

* Correspondence: skontog@uoi.gr; Tel.: +30-26510-08252

Abstract: This paper presents a bee-condition-monitoring system incorporated with a deep-learning process to detect bee swarming. This system includes easy-to-use image acquisition and various end node approaches for either on-site or cloud-based mechanisms. This system also incorporates a new smart CNN engine called Swarm-engine for detecting bees and the issue of notifications in cases of bee swarming conditions to the apiarists. First, this paper presents the authors' proposed implementation system architecture and end node versions that put it to the test. Then, several pre-trained networks of the authors' proposed CNN Swarm-engine were also validated to detect bee-clustering events that may lead to swarming. Finally, their accuracy and performance towards detection were evaluated using both cloud cores and embedded ARM devices on parts of the system's different end-node implementations.

Keywords: beehive-monitoring systems; IoT; convolutional neural networks; image processing; performance evaluation; distributed systems

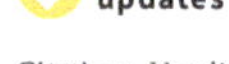

Citation: Voudiotis, G.; Kontogiannis, S.; Pikridas, C. Proposed Smart Monitoring System for the Detection of Bee Swarming. *Inventions* **2021**, *6*, 87. https://doi.org/10.3390/inventions6040087

Academic Editors: Francisco Manzano Agugliaro and Esther Salmerón-Manzano

Received: 30 October 2021
Accepted: 15 November 2021
Published: 17 November 2021

1. Introduction

The Internet of Things (IoT) industry is shifting fast towards the agricultural sector, aiming for the vast applicability of new technologies. Existing applications in agriculture include environmental monitoring of open field agricultural systems, the food supply chain monitoring, and livestock monitoring [1–4]. Several bee-monitoring and beekeeping-resource management systems or frameworks that incorporate IoT and smart services have been proposed in the literature [5–7], while others exist as market solutions. This paper investigates existing technological systems focusing on detecting bee stress, queen succession, or Colony Collapse Disorder (CCD), favorable conditions that can lead to bee swarming. Swarming is when honeybee colonies reproduce to form new ones or when a honeybee colony becomes too congested or stressed and requires beekeeping treatments.

Swarming is the phenomenon of bee clustering that indicates a crowded beehive that usually appears under normal conditions. At its first development stages, it seems at the end frame of the brood box or in the available space between the upper part of the frames and the beehive lid.

In most cases, bee swarming is a natural phenomenon that beekeepers are called to mitigate with new frames or floors. Nevertheless, in many cases, bee clustering events that lead to bee swarming may occur in cases such as: (a) Varroa mite disease outbursts that lead to the replacement of the queen, (b) the birth of a new queen, which takes part of the colony and abandons the beehive, or (c) extreme environmental conditions or even low pollen supplies, which decrease the queen's laying and force her to migrate. All of the above cases (a),(b), and (c) lead to swarming events. The following paragraphs describe how the variation of the condition parameters can lead to swarming events.

The environment inside and around the beehives is vital to the colony establishment's success and development. An essential factor in apiary hives that affects both colony survival and honey yield is the ability to manage agricultural interventions and disease treatments (especially Varroa mite [8]) and monitor the conditions inside the beehive [9–11].

Beehive colonies are especially vulnerable to temperature fluctuations, which in turn affect honey yields. At high temperatures above 35 °C, honeybees are actively involved in the thermoregulation of the colony with fanning activities. For low temperatures below 15 °C, they reduce their mobility and gather at the center of the brood box, forming bee clusters and consuming collected honey [5,12]. Regarding the air temperature inside the beehives, a mathematical model was presented in [13]. Moreover, bee swarming is also strongly correlated with temperature variations [14]. Therefore, to alleviate temperature stress events, inexpensive autonomous temperature-monitoring devices [15] and IoT low-power systems have been proposed in the literature. Such systems are the WBee system [16], an RF temperature monitoring system with GPS capabilities presented in [14], the BeeQ Resources Management System [17], and the SBMaCS, which utilizes piezoelectric-energy-harvesting functionalities for temperature measurements' data transmissions [18].

Similarly, high humidity levels can lead to swarming events inside the beehive, excessive colony honey consumption, and the production of propolis and wax by the bees as a countermeasure [19,20]. Beehive humidity levels' reduction concerning atmospheric humidity is achieved with the opening/closing of the ventilation holes attached to the beehive lid [21], or with the use of automated blower fans [22], or integrated Peltier systems [12].

The use of weight scales with the adoption of resource-monitoring systems and treatment protocols may also be of assistance to mitigate swarming events based on colony strength, and constitution [23]. It has been reported that significant weight reduction for a fully deployed ten-frame beehive cell below 19 kg indicates swarming in progress requiring immediate attention and feeding interventions, as confirmed by apiarists.

Apart from temperature, humidity, and weight monitoring, sound monitoring has also been proposed to detect swarming events, as provided by the analysis in [24]. The authors in [25] classified the swarming spectrum at 400–500 Hz with a duration interval of at least 35 min. Using Mel Frequency Cepstral Coefficients (MFCCs) and deltas, as well as Mel band spectra computational sound scene analysis, the authors in [26], tried to detect coefficients that denote swarming. The authors in [27] used MFCC feature coefficient vectors and trained a system using the Gaussian Mixture Model (GMM) and Hidden Markov Model (HMM) classifiers to detect swarming. Furthermore, the authors in [28] proposed a support vector machine classifier of MFCC coefficients and a CNN model that can also be used for swarm detection.

Beehive cameras in beekeeping have been used mainly as a security and antitheft protection instrument rather than a condition-monitoring one. Nevertheless, significant steps over the last few years have been taken, using mainly image-processing techniques, circular Hough transformations [29], background subtractions, and eclipse approximations, to detect bees or bee flying paths at the hive entrances [30]. Apart from image processing, another system used for bee counting using IR sensors at the beehive entrances was presented in [31]. Image processing is the easiest to implement and most reliable source for swarming or external attack alert detection. Nevertheless, if performed at the hive entrance, it cannot reliably detect the occurring event early enough for the beekeepers to ameliorate it successfully.

Since swarming events occur mainly during the day and in the spring and summer months, as indicated by apiarists, this paper presents a new camera sensor system for detecting swarming inside the beehive. The proposed system uses a camera that incorporates an image-processing motion logic and utilizes a pre-trained Convolutional Neural Network (CNN) to detect bees. The rest of this paper is structured as follows: Section 2 presents related work in existing beehive-condition-monitoring products on the market. Section 3 presents the authors' proposed monitoring system and the system's capabilities. Section 4 presents the CNN algorithms used by the system and detection algorithm process. Section 5 presents the authors' experimentation on different end node devices, CNN algorithms and models as well as real testing for bees detection and bee swarming. Finally, the paper summarizes the findings and experimental results of the system.

2. Related Work on Beehive-Condition-Monitoring Products

In this section, the relevant research is presented based on which devices have been used for monitoring and recording the bee swarming conditions that prevail at any given time inside the beehive cells. The corresponding swarming-detection services provided by each device are divided into three categories: (a) honey productivity monitoring using weight scales, image motion detection, or other sensors, (b) direct population monitoring using cameras and smart AI algorithms and deep neural networks, (c) indirect population monitoring only using audio, or other sensors, or cameras that are not located externally, or cameras without any smart AI algorithm for automated detection. The prominent representatives of the devices performing the population monitoring are the Bee-Shop Camera Kit [32] and the EyeSon [33] for Category (a) and the Zygi [34], Arnia [35], Hive-Tech [36], and HiveMind [37] devices for Category (c). On the other hand, no known devices on the market directly monitor beehives using cameras inside the beehive box (Category (b)). Table 1 summarizes existing beehive-monitoring systems concerning population monitoring and productivity.

Table 1. Existing bee-swarming-monitoring systems.

System/Device	Swarm-Monitoring Capabilities		
	Productivity Monitoring	Indirect Population Monitoring Using Cameras	Indirect Population Monitoring Using Sound or Other Sensors
Bee-Shop camera kit [32]	✓	✓	
HiveMind system [37]	✓		✓
Hive-Tech system [36]	✓		✓
Arnia system [35]	✓		✓
EyeSon Hive system [33]	✓	✓	
Zygi [34]	✓	✓	

The Bee-Shop [32] monitoring equipment can observe the hive's productivity through video recording or photos. The monitoring device is placed in front of the beehive door. The captured material is stored on an SD card. It can be sent to the beekeeper's mobile phone using the 3G/4G LTE network, showing the contours of bee swarms as detected from the image-detection algorithm included in the camera kit. Similarly, the Bee-Shop camera kit offers motion detection and security instances for the apiary.

Similarly, EyeSon Hives [33] uses an external bee box camera and an image-detection algorithm to record the swarms of bees located outside the hive and algorithmically analyze the swarm flight direction. EyeS on Hives [33] also uses 3G/4G LTE connectivity and enables the beekeeper to stream video via a mobile phone application in real-time.

Zygi [34] provides access to weight measurements. It is also capable of a variety of external measurements such as temperature and humidity. Nevertheless, since this is performed externally, such weight measurements do not indicate the conditions that apply inside the beehive box. Zygi also includes an external camera placed in front of the bee box and transmits photo snapshots via GSM or GPRS. However, this functionality does not have a smart engine or image-detection algorithm to detect swarming and requires beekeepers' evaluation.

Devices similar to Zygi [34] are the Arnia [35], Hive-Tech [36], and HiveMind [37] devices, which are assumed to be indirect monitoring devices due to the absence of a camera module.

Hive-Tech [36] can monitor the bee box and detect swarming by using a monitoring IR sensor or reflectance sensors that detect real-time crowd conditions at the bee box openings (where the sensors are placed), as well as bee mobility and counting [38,39]. The algorithm used is relatively easy to implement, and swarming results can be derived using data analysis.

Arnia [35] includes a microphone with audio-recording capabilities, FFT frequency spikes, Mel Frequency Cepstral Coefficient (MFCC) deltas' monitoring [28], and notifications. Finally, HiveMind [37] includes humidity and temperature sensors and a bee activity sound/IR sensor as a good indicator for overall beehive doorway activity.

3. Proposed Monitoring System

The authors propose a new incident-response system for automatic detection of swarming. The system includes the following components: (1) The Beehive-Monitoring Node, (2) The Quality Resource Management System and (3) The Beehive end node application. System components are described in the subsections that follow.

3.1. Beehive-Monitoring Node

The beehive-camera-monitoring node is placed inside the beehive's brood box and includes the following components:

The camera module component. The camera module component is responsible for the acquisition of still images inside the beehive. It is placed on a plastic frame covered entirely with a smooth plastic surface to avoid being built on or waxed by bees. The camera used is a fish-eye lens of a 180–200° view angle, a 5MPixel camera with LEDs included (brightness controlled by the MCU), and adjustable focus distance connected directly to the ARM using the MIPI CSI/2 interface using a 15 pin FFC cable. The camera module can take ultra-wide HD images of 2592 × 1944 resolution and achieve frame rates of 2 frames/s for single-core ARM, 8 frames/s for quad-core ARM, and 23 frames/s for octa-core ARM devices, at its maximum resolution potential

The microprocessor control unit, responsible for storing camera snapshots and uploading them to the cloud (for Version 1 end node devices) or responsible for taking camera snapshots and implementing the deep-learning detection algorithm (for Version 2 end node devices)

The network transponder device, which can be either a UART-connected WiFi transponder (for Version 1 nodes) or an ARM microprocessor, including an SPI connected LoRaWAN transponder (for Version 2 nodes)

The power component, which includes a 20 W/12 V PV panel connected directly to a 12 V-9 Ah lead-acid SLA/AGM battery. The battery is placed under the PV panel on top of the beehive and feeds the ARM MCU unit using a 12-5 V/2 A buck converter. The battery used is a deep depletion one since the system might, due to its small battery capacity, be fully discharged, especially at night or on prolonged cloudy days

3.2. Beehive Concentrator

The concentrator is responsible for the nodes' data transmission over the Internet to the central storage and management unit, called the Bee Quality Resource Management System. It acts as an intermediate gateway among the end nodes and the BeeQ RMS application service and web interface (see Figure A4). Depending on the version of the nodes, for node v1, the beehive concentrator is a Wi-Fi access point over the 3G/4G LTE cellular network, and Version 2 is a LoRaWAN gateway over 3G/4G LTE. Figure 1a,b illustrate the Node-1 and Node-2 devices and their connectivity to the RMS over the different types of beehive concentrators. The technical specifications and capabilities also differ among the two implementation versions.

The Version 1 node concentrator (see Figure 1) can upload images with an overall bandwidth capability that varies from 1–7/10–57 Mbps, depending on the gateway's distance from the beehive and limited by the LTE technology used. Nevertheless, it is characterized as a close-distance concentrator solution since the concentrator must be inside the beehive array and at a LOS distance of no more than 100 m from the hive. The other problem with Version 1 node concentrators is that their continuous operation and control signaling transmissions waste 40–75% more energy than LoRaWAN [40].

For Version 2 devices, the LoRaWAN concentrator is permanently set to a listening state and can be emulated as a class-C single-channel device (see Figure 2). In such cases, for a 6 Ah battery, which can deliver all its potential, the expected gateway uptime is 35–40 h [41] (without calculating the concentrator's LTE transponder energy consumption). The coverage distance of the Version 2 concentrator also varies since it can cover distances up to 12–18 km for Line Of Sight (LOS) setups and 1–5 km for non-LOS ones [42]. Furthermore, its scalability differs since it can deliver at least 100–250 nodes per concentrator at SF-12 and a worst-case packet loss of 10–25% [43], concerning a maximum of 5–10 nodes for the WiFi ones. The disadvantage of the Class-2 node is that its Bandwidth (BW) potential is limited to 0.3–5.4 Kbps, including the duty cycle transmission limitations [44]:

3.3. Beehive Quality Resource Management System

The BeeQ RMS system is a SaaS cloud service capable of interacting with the end nodes via the concentrator and the end-users. For Version 1 devices, the end nodes periodically deliver images to the BeeQ RMS using HTTP put requests (Figure 1 P1). Then, the uploaded photos are processed at the RMS end using the motion detection and CNN algorithm and web interface (see Figure A5) of the BeeQ RMS swarming service (Figure 1 P2).

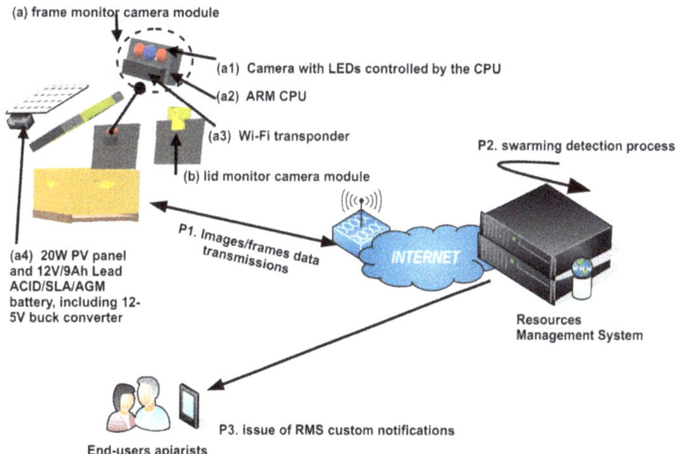

Figure 1. Beehive monitoring end node Version 1 device that includes a WiFi transponder and the beehive-monitoring system logic using v1 end node devices.

For Version 2 devices, the motion detection and CNN bee detection algorithm are performed directly at the end node. When the detection period is reached and only when bee motion is detected, the trained CNN engine is loaded, and the number of bees is calculated, as well as the severity of the event. The interdetection interval is usually statically set to 1–2 h. However, for Version 2 devices, the detection outcome is transmitted over the LoRaWAN network. It is collected and AES-128 decrypted by the BeeQ RMS LoRa application server, sending the detection message over MQTT via the BeeQ RMS MQTT broker. The BeeQ RMS MQTT then stores the message in the DB service, where the MQTT message is JSON decoded and stored in the BeeQ RMS MySQL database (see Figure 2, P1 and P2) [4]. Similarly, for Version 1 devices, the images are processed by the BeeQ RMS Swarm detection service (Figure 1 P2), responsible for the swarm detection process and storing the detection result at the BeeQ RMS MySQL database.

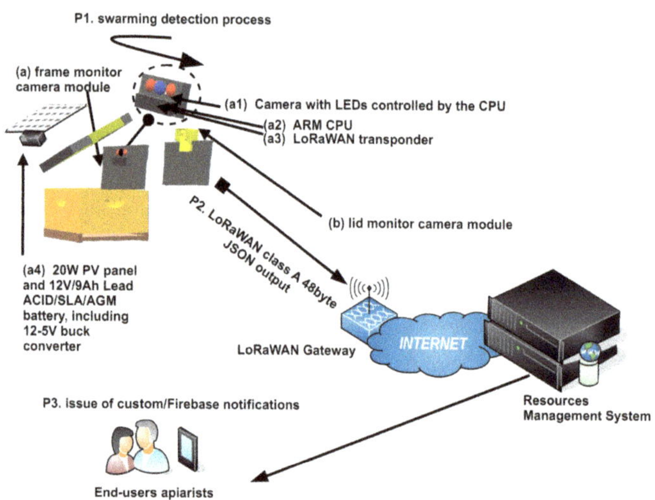

Figure 2. Beehive monitoring end node Version 2 device that includes a LoRaWAN transponder and beehive-monitoring system logic using v2 end node devices.

For both cases (Version 1, Version 2), the output result (number of detected bees and severity of alert) is delivered to the end-users using push notifications. The Bee-RMS mobile phone application and dashboard are notified accordingly (Figure 1 P3 and Figure 2 P3). The Version 1 and Version 2 device prototypes are illustrated in Figure A6.

3.4. *Beehive End Node Application*

The end node applications used by the BeeQ RMS system are the Android mobile phone app and the web panel. Both applications share the same operational and functional characteristics, that is, recording feeds and periodic farming checks, sensory input feedback for temperature, sound level increase, or humidity-related incidents (sensors fragment and web panels), and swarm detection alerts via the proposed system camera module. In addition, the Firebase push notifications service is used [45], while for the web dashboard, the jQuery notify capabilities are exploited.

4. Deep-Learning System Training and Proposed Detection Process

In this section, the authors describe their utilized deep-learning detection process for bee swarming inside the beehive. The defined approach is a part of the pre-trained CNN models, and CNN algorithms [46,47], which after the use of motion-detection filtering, try to estimate if swarming conditions have been reached.

The process used to build and test the swarming operation for detecting bees includes four steps for the CNN-training process. Furthermore, the detection service also consists of three stages, as illustrated in Figure 3. The CNN training steps used are as follows:

Step 1—Initial data acquisition and data cleansing: The initial imagery dataset acquired by the beehive-monitoring module is manually analyzed and filtered to eliminate blurred images or images with a low resolution and light intensity. The photos in this experimentation taken from the camera module are set to a minimum acquisition of 0.5 Mpx in size of 800 $\times$ 600 px 300 dpi (67.7 $\times$ 50.8 mm^2) compressed in the JPEG format using a compression ratio Q = 70 (22,91) of 200–250 KB each. That is because the authors wanted to experiment with the smallest possible size of image transmission (due to the per GB network provider costs of image transmissions for Version 1 devices or to minimize processing time overheads for Version 2 devices). Similarly, the trained CNN and algorithms used are the most processing-light for portable devices, using a minimum trained image size input of 640 × 640 px (lightly distorted at the image height) and using cubic interpolation.

The trained Convolutional Neural Network (CNN) is used to solve the problem of swarming by counting the bee concentration above the bee frames and inside the beehive lid. The detection categories that the authors' classifier uses are:

Class 0: no bees detected;
Class 1: a limited number of bees scattered on the frame or the lid (less than 10);
Class 2: a small number of bees (less than or equal to 20);
Class 3: initial swarm concentration and a medium number of bees concentrated (more than 20 and less than or equal to 50);
Class 4: swarming incident (high number of bees) (more than 50).

For each class, the number of detected bees was set as a class identifier (the class identifier boundaries can be arbitrarily set accordingly to the detection service configuration file). Therefore, the selected initial data-set may consist of at least 1000 images per detection class, a total of 5000 images used for training the CNN;

Step 2—Image transformation and data annotation: The number of collected images per class used for training was annotated by hand using the LabelImg tool [48]. Other commonly used annotation tools are Labelbox [49], ImgAnnotation [50], and Computer Vision Annotation Tool [51], all of which provide an XML annotated output.

Image clearness and resolution are equally important in the case of initially having different image dimensions. Regarding photo clearness, the method used is as follows. A bilateral filter smooths all images using a degree of smoothing sigma = 0.5–0.8 and a small 7 × 7 kernel. Afterward, all photos must be scaled to particular and fixed dimensions to be inserted into the training network. Scaling is performed either using a cubic interpolation process, or a super-resolution EDSR process [52]. The preparation of the images is based on the dimensions required as the input by the selected training algorithm, which matches with the underlying extent of the initial CNN layer. The image transformation processes were implemented using OpenCV [53] and were also part of the 2nd stage of the detection process (detection service) before their input into the CNN engine (see Figure 3);

Step 3—Training process: The preparation of the training process is based on using pre-trained Convolution Neural Network (CNN) models [54,55], TensorFlow [56] (Version 1), and the use of all available CPU and GPU system resources. To achieve training parallel execution speed up, the use of a GPU is necessary, as well as the installation of the CUDA toolkit such that the training process utilizes the GPU resources, according to the TensorFlow requirements [57].

The CNN model's design includes selecting one of the existing pre-trained TensorFlow models, where our swarming classifier will be included as the final classification step. Selected core models for TensorFlow used for training our swarming model and their capabilities are presented in Table 2. Once the Step 2 annotation process is complete and the pre-trained CNN model is selected, the images are randomly divided into two sets. The training set consisted of 80% of the annotated images, and the testing set contained the remaining 20%. The validation set was also used by randomly taking 20% of the training set;

Step 4—Detection service process: This process is performed by a detection application installed as a service that loads the CNN inference graph in memory and processes arbitrary images received via HTTP put requests from the node Version 1 device. The HTTP put method requires that the requested URI be updated or created to be enclosed in the put message body. Thus, if a resource exists at that URI, the message body should be considered as a new modified version of that resource. If the put request is received, the service initiates the detection process and, from the detection JSON output for that resource, creates a new resource XML response record. Due to the asynchronous nature of the swarming service, the request is also recorded into the BeeQ RMS database to be accessible by the BeeQ RMS web panel and mobile phone application. Moreover, the JSON output response, when generated, is also pushed to the Firebase service to be sent as push notifications to the BeeQ RMS mobile phone application [45,58]. Figure 3 analytically illustrates the detection process steps for both Version 1 and Version 2 end nodes.

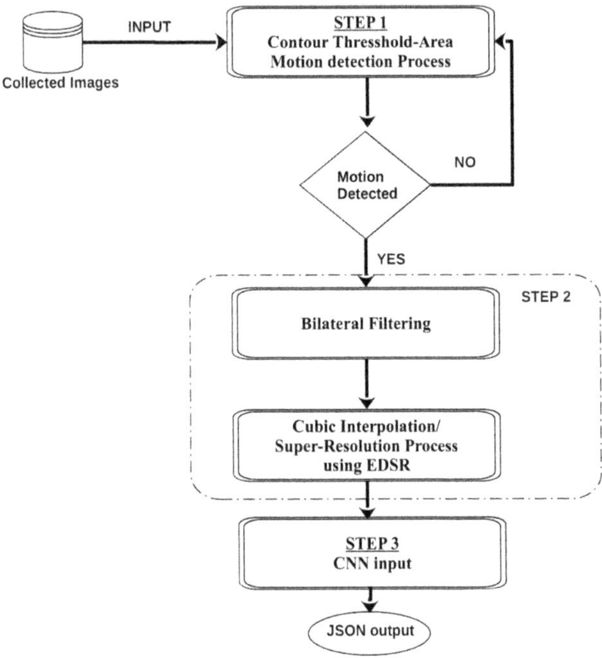

Figure 3. Bee swarming service automated detection process.

Step 1 is the threshold max-contour image selection process issued only by Version 2 devices as part of the sequential frames' motion detection process instantiated periodically. Upon motion detection, photo frames that include the activity notification contours for Version 2 devices or uploaded frames from Version 1 devices are transformed through a bilateral filtering transformation with sigma_space and sigma_color parameters equal to 0.75 and a pixel neighborhood of 5 px. Upon bilateral filter smoothing, the scaling process initiates to normalize images to the input dimensions of the CNN. For down-scaling or minimum dimension up-scaling (up to 100 px), cubic interpolation is used, while for large up-scales, the OpenCV super-resolution process is instantiated (Enhanced Deep Residual Networks for Single Image Super-Resolution) [52] using the end node device. Upon CNN image normalization, the photos are fed to the convolutional neural network classifier, which detects the number of bee contours and reports it using XML output image reports, as presented below:

```
<detection>
<bees>Number of bees</bees>
<varroa>
<detected>True/False</detected>
<num>Number of bees</num>
</varroa>
<queen>
<detected>True/False</detected>
</queen>
<hornets>
<detected>True/False</detected>
<num>Number of hornets</num>
</hornets>
<notes>
Instructions for dealing with diseases.
</notes>
</detection>
```

Apart from bee counting information, the XML image reports also include the information of detected bees carrying the Varroa mite. Such detection can be performed using two RGB color masks over the detected bee contours. This functionality is still under validation and therefore set as future work and exploitation of the CNN bee counting classifier. Nevertheless, this capability was included in the web interface but not thoroughly tested. Similarly, the automated queen detection functionality (now performed using the check status form report in the Bee RMS web application) was also included as a capability of the web detection interface. The Bee RMS classifier is still under data collection and training since the preliminary trained models used a limited number of images. The algorithmic process includes a new detection queen bee class and the HSV processing of the bee queen's color to estimate its age.

Upon generation of the XML image report, the swarming service loads the report and stores it in the BeeQ RMS database to be accessible by the web BeeQ RMS interface and transforms it to a JSON object to be sent to the Firebase service [45,58]. The BeeQ RMS Android mobile application can receive such push notifications and appropriately notify the beekeeper.

The following section presents the authors' experimentation using different trained neural networks and end node devices.

Table 2. CNN models used during the training step.

CNN Mode [54,55]	CNN Models Capabilities			
	Input Images Size (wxh) (px)	**Model Classes**	**Models Mean Detection Time per (ms)**	**Model Accuracy mAP**
SSD - MobileNet v1	300 × 300	90	30	21
SSD - MobileNet v2	300 × 300	90	31	22
SSD - Inception v2	300 × 300	90	42	24
Faster-RCNN - Inception v2	600–1024 × 600–1024	90	58	28

5. Experimental Scenarios

In this section, the authors present their experimental scenarios using Version 1 and Version 2 end node devices and their experimental results while detecting bees using the two selected CNN algorithms (SSD and Faster-RCNN) and two selected models: MobileNet v1 and Inception v2. These models were chosen from a set of pre-trained models based on their low mean detection time and good mean Average Precision (mAP) results (as illustrated in Table 2). Two different CNN algorithms were used during experimentation; SSD and Faster R-CNN. Using the SSD algorithm, two different pre-trained models have been utilized: The MobileNet v1 and Inception v2 COCO models. For the Faster R-CNN algorithm the Inception v2 model has been used.

For the CNN training process, the total number of images in the data-set annotated and used for the models' training in this scenario was 6627 with dimensions of 800 $\times$ 600 px. One hundred of them were randomly selected for the detection validation process. The constructed network was trained to detect bee classes, as described in the previous section. The Intersection over Union (IoU) threshold was set to 0.5.

During the training processes, the TensorBoard has been used. The TensorBoard presents and records the loss charts during training. The values shown in the loss diagrams are the values obtained from the loss functions of each algorithm [59]. Looking at the recorded loss charts, it is evident whether the model would have a high degree of accuracy in detection. For example, if the last calculated loss is close to zero, it is expected for the model to have a high degree of accuracy. In contrast, the further away from zero, the more the accuracy of the model decreases. After completing the training of the three trained networks, the initial and final loss training values and the total training time are presented in Table 3.

The authors performed their validation detection tests using 100 beehive photos. For each test, five different metrics were measured: the time required to load the trained network into the system memory, the total detection time, the average ROI detection time, the mean Average Precision (mAP) derived from the models' testing for IoU = 0.5, and the maximum memory allocation per neural network.

Furthermore, the detection accuracy (DA) and the mean detection accuracy (MDA) were also defined by the authors. That is, by performing manual bee counting for each detected bee contour with a confidence level threshold above 0.2 over the total images. The detection accuracy (DA) and the mean detection accuracy (MDA) metrics are calculated using the Equation (1):

$$DA = \frac{DC}{N}, MDA = \frac{1}{M}\sum_{i=1}^{M} DA_i \tag{1}$$

where variable DC is the contours that have been marked and have bees, the variable N is the total number of bees in the photo, while the variable M is the number of photos we tested. The following error metric (Er) was also defined, which occurs during the detection process, using the Equation (2):

$$Er = 1 - \frac{DC}{C} \tag{2}$$

where variable C is the total number of contours marked by the model. The accuracy of the object detection was also measured by the mAP using TensorBoard. That is, the average of the maximum precisions at different detected contours over the real annotated ones. Precision measures the prediction accuracy by comparing the true positives and all the false-negative cases.

The tests have been performed on three different CPU Version 2 devices and in the cloud by utilizing either a single-core cloud CPU or a 24-core cloud CPU for testing Version 1 end node devices' cloud processing requirements. The memory capabilities for the single-core cloud CPU were 8 GB, while for the 24-core cloud CPU, 64 GB. For the Version 2 tested devices, the authors utilized a Raspberry Pi-3 armv7 1.2 GHz quad-core with 1 GB of RAM and a 2 GB swap, and a Raspberry Pi zero armv6 1GHz single-core, 512 MB RAM, and

2 GB swap. Both RPi nodes' OS was the 32 bit Raspbian GNU/Linux operating system. A third Version 2 end node device was also tested the NVIDIA Jetson Nano board. Jetson has a 1.4 GHz quad-core armv7 processor, 128-core NVIDIA Maxwell GPU, and its memory capacity is 4 GB, used either by the CPU or GPU. The Jetson operating system is a 32 bit Linux for Tegra.

Table 3. CNN trained models' loss values and total training time.

CNN Models	Initial Loss Value	Final Loss Value	Total Training Time
SSD - MobileNet v1	11.90	0.7025	35 h 47 min
SSD - Inception v2	4.095	0.8837	43 h 15 min
Faster-RCNN - Inception v2	1.438	0.0047	14 h 27 min

5.1. Scenario I: End Node Version 1 Detection Systems' Performance Tests

Scenario I detection tests included Version 1 end node device-equipped systems' performance test using cloud (a) single-core and (b) multi-core CPUs and Version 1 end node devices. During these tests for the (a) and (b) system cases, the two selected algorithms (SSD, Faster-RCNN) and their trained models were tested for their performance. The results for Cases (a) and (b) are presented in Tables 4 and 5 accordingly.

Table 4. Experimental Scenario I, x64 CPU CNN detection time and memory usage.

CNN Models	Load Time (s)	Testing Detection Time (s)	Detection Average Time per Image (s)	Total Time (s)	Memory Usage (MB)
SSD - MobileNet v1	0.291	1.340	0.105	1.631	45.262
SSD - Inception v2	0.432	1.523	0.156	1.955	106.889
Faster-RCNN - Inception v2	0.368	5.358	1.079	5.726	104.327

Based on Scenario I's result for single-core x86 64-bit CPUs, using the SSD on the MobileNet v1 network provided the best TensorFlow network load time (32% less than the SSD Inception model and 15% less than the Faster-RCNN Inception model). Comparing the load times of the Inception models for the SSD and Faster-RCNN algorithms showed that Faster-RCNN provided an optimum faster loading model than its SSD counterpart. Nevertheless, the SSD single-image average detection time was 2.5-times faster than its Faster-RCNN counterpart. Similar results were applied and for SSD-MobileNet v1 (three times faster).

Comparing Tables 3 and 4, it is evident that Faster-RCNN Inception v2 had the minimum total loss value (0.0047%), as indicated by the training process and the least training time, while the SSD algorithms showed high loss values of 0.7% and 0.8% accordingly. Having as the performance indicator metric the Mean Detection Accuracy (MDA), provided in Table 7, over the total detection time per image, the authors defined a CNN evaluation metric called the Success Frequency (S.F.) metric, expressed by Equation (3). However, if it is not possible to validate the model using the MDA metric, the model mAP accuracy values can be used instead (see Equation (4)).

$$SF = \frac{MDA}{T_{load} + T_{detect}} \quad (3)$$

$$SF_{mAP} = \frac{mAP}{T_{load} + T_{detect}} \quad (4)$$

where T_{load} is the mean frame load time (s) and T_{detect} is the mean CNN frame detection time (s). The S.F. metric expresses the number of ROIs (contours) per second successfully detected over time. The S.F. values over the mAP and image detection times are depicted in Figure 4. The S.F. metric is critical for embedded and low-power devices with limited processing capabilities (studied in Scenario II). Therefore, it is more suitable for the CNN algorithm to be selected based on the highest S.F. value instead of the CNN mAP or total loss for such devices.

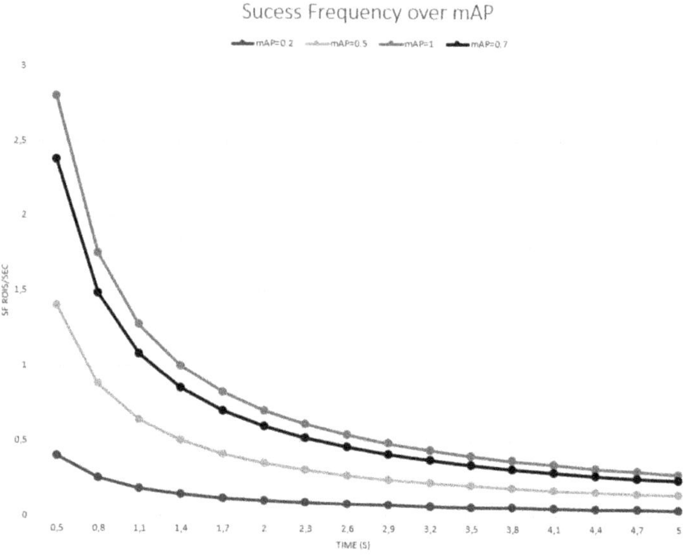

Figure 4. S.F. metric in ROIs/second over the mAP.

Comparing the single x86 64bit CPU measurements with the 24-CPU measurements (Tables 4 and 5), the speedup value σ can be calculated using the mean detection time as $\sigma = \frac{T_1}{T_{24}}$ since the mean CNN detection time is a parallel task among the 24 cores. Therefore, the speedup achieved using 24 cores for the SSD algorithm was almost constant, close to $\sigma = 1.1$. Therefore, with the SSD algorithm for cloud 24-core CPUs, using a single-core provided the same results in terms of performance. However, for the Faster-RCNN, using 24 cores offered a double performance speedup of $\sigma = 2.3$, that is to reduce the per-image detection time, 50%, at least 24 cores would need to operate in parallel, contributing to the detection process of Faster-RCNN.

5.2. Scenario II: End Node Version 2 Detection Systems' Performance Tests

Scenario II detection tests included Version 2 devices' performance tests as standalone devices (no cloud support). The end node systems that were tested were: (a) single-core ARMv6, (b) quad-core ARMv7, and (c) CPU+GPU quad-core ARMv7 Jetson device. During these tests, the two selected algorithms (SSD, Faster-RCNN) and their trained models were tested in terms of performance. The performance results are presented in Table 6.

Table 5. Experimental Scenario I, x64, 24-core CPU detection time and memory usage.

CNN Models	Load Time (s)	Testing Detection Time (s)	Detection Average Time per Image (s)	Total Time (s)	Memory Usage (MB)
SSD - MobileNet v1	0.309	1.335	0.087	1.644	45.262
SSD - Inception v2	0.431	1.361	0.109	1.792	106.889
Faster-RCNN - Inception v2	0.386	2.249	0.278	2.635	104.327

Table 6. Experimental Scenario II, RPi ARMv7, ARMv6 single-core and Jetson Nano CPU ARMv7 + GPU performance comparison results.

CNN Models	Load Time (s)	Mean CNN Time (s)	Mean ROI Time (s)	Total Time (s)	MeM Usage (MB)	S.F. mAP
	ARM single-core	ARM single-core	ARM single-core	ARM single-core	ARM single-core	ARM single-core
	ARM quad-core	ARM quad-core	ARM quad-core	ARM quad-core	ARM quad-core	ARM quad-core
	ARM quad-core + GPU	ARM quad-core + GPU	ARM quad-core + GPU	ARM quad-core + GPU	ARM quad-core + GPU	ARM quad-core + GPU
SSD - MobileNet v1	10.97	29.297	2.955	40.267	75.924	0.01
	10.597	22.561	2.457	33.158	75.924	0.012
	7.916	33.524	6.606	41.44	81.02	0.01
SSD - Inception v2	15.381	34.884	6.614	50.265	174.449	0.008
	15.318	36.072	6.886	51.39	174.449	0.0079
	10.614	61.106	13.416	71.72	181.08	0.0056
Faster-RCNN - Inception v2	14.389	184.373	41.826	198.762	166.351	0.003
	14.243	111.333	24.222	125.576	166.351	0.0058
	9.957	69.968	15.5	79.926	172.729	0.009

As shown in Table 6, the S.F. mAP values for the embedded micro-devices indicated that the best algorithm to use was the SSD with the MobileNet v1 network. This algorithm had similar S.F. value results for single-core ARM, quad-core, and Jetson devices. For less accurate detection networks of mAP values less than 0.5, there were no significant gains from using multi-core embedded systems.

This was not the case for high-accuracy detection devices, which can provide mAP values of more than 0.7, such as the Faster-RCNN. In these cases, the use of multiple CPUs and GPUs can offer significant gains of 40–50%, in terms of S.F. (both detection time reduction and accuracy increase as signified by the mAP). Since energy requirements for the devices are critical, for low-energy devices, the SSD MobileNet v1 is preferred using an ARMv6 single-core CPU (since the S.F. increase of ARMv6 concerning the use of four cores to achieve such a result was considered by the authors to be a significant energy expenditure).

For high-accuracy devices, the Jetson board using the Faster-RCNN algorithm provided the best results in terms of accuracy and execution time. Performance tests on the Jetson Nano microcomputer showed that the results obtained from this system for high accuracy were better than the ones from the RPi 3, due to the GPU's participation in the detection process. However, in the Jetson Nano microprocessor, some transient errors occurred during the CPU and GPU allocation, which did not cause significant problems during the detection process. Nevertheless, since no energy measurements were performed for the quad-core RPi and Jetson, the RPi can also be considered a low-energy, high-accuracy alternative instead of the Jetson Nano, according to the devices' data-sheets.

5.3. Scenario III: CNN Algorithms' and Models' Accuracy

Scenario III's detection tests focused on the accuracy of the two used algorithms and their produced trained CNNs, using the mean detection accuracy metric from Equation (1) and the mAP values calculated by the TensorBoard during the training process. The results are presented in Table 7. Furthermore, the detection image results are illustrated in Figures A1–A3 in Appendix A, using the Jetson Nano Version 2 device and the SSD and Faster-RCNN algorithms on their trained models.

Table 7. Mean detection accuracy (manual bee counting verification) over models' mAP.

CNN Models	Mean Detection Accuracy (MDA)	CNN Model mAP	x86-64 Single-Core CPU SF	ARM Single-Core CPU SF
SSD - MobileNet v1	0.418	0.4223	0.009	0.01
SSD - Inception v2	0.13	0.4083	0.001	0.002
Faster-RCNN - Inception v2	0.703	0.7308	0.006	0.003

According to the models' accuracy tests, the following conclusions were derived. First, based on the training process (Table 3), the Faster-RCNN algorithm with the Inception v2 model was faster to train than SSD. In addition, based on Tables 3 and 7, the authors concluded that the lower the values of the training losses, the better the results we would obtain during the detection. It is also apparent that the best model for Version 2 devices with limited resources is SSD-MobileNet v1, as also shown in Table 7, by the S.F. values. That is because it requires less memory and processing time to work; regardless, it was 30% less accurate than the Faster-RCNN algorithm in terms of the MDA metric.

5.4. Scenario IV: System Validation for Swarming

The proposed swarming detection system was validated for swarming in two distinct cases:

Case 1—Overpopulated beehive. In this test case, a small beehive of five frames was used and monitored for a period of two months (April 2021–May 2021), using the Version 1 device. The camera module was placed (see Figure A6A, end node Version 1), facing the last empty beehive frame. The system successfully managed to capture the population increase (see Figure A5), reaching from the detection of Class-1 to a Class-3 initial swarm concentration and a medium number of bees concentrated. As a provocative measure, a new frame was added.

Case 2—Provoked swarming. In this test case, in a beehive colony during the early spring periodic check (March 2021), a new bee queen cell was detected by the apiarists, indicating the incubation of a new queen. The Version 1 device with the camera module was placed facing the area above the frames, towards the ventilation holes. Between the frames and the lid, the beehive's progress was monitored weekly (weekly apiary checks). In the first two weeks, a significant increase of bee clustering was monitored, varying from the detection of Class-1 to Class-3 and back to Class-0. The apiarists also matched the swarming indication to the imminent swarming event since a significant portion of the hive population had abandoned the beehive.

In the above-mentioned cases, Case-1 experiments were successfully performed more than once. Both validation test cases were performed at our laboratory beehive station located in Ligopsa, Epirus, Greece, and are mentioned in this section as proof of the authors' proposed concept. Nevertheless, more extensive validation and evaluation are set as future work.

6. Conclusions

This paper presented a new beekeeping-condition-monitoring system for the detection of bee swarming. The proposed system included two different versions of the end node devices and a new algorithm that utilizes CNN deep-learning networks. The proposed algorithm can be incorporated either at the cloud or at the end node devices by modifying the system's architecture accordingly.

Based on their system end-node implementations, the authors' experimentation focused on using different CNN algorithms and end node embedded modules. The authors also proposed two new metrics, the mean detection accuracy, and success frequency. These metrics were used to verify the mAP and total loss measurements and to make a trade-off between detection accuracy and limited resources due to the low energy consumption requirements of the end node devices.

The authors' experimentation made it clear that low-processing mobile devices can use less accurate CNN detection algorithms. Instead, metrics can be used that can accurately represent each CNN's resource utilization, such as S.F., MDA, and speedup (σ), for the acquisition of either the appropriate embedded device configuration or cloud resource utilization. Furthermore, appropriate validation of the authors' proposed system was performed in two separate cases of (a) bee queen removal and (b) a beehive population increase that may lead to a swarming incident if not properly ameliorated with the addition of frames in the beehive.

The authors set at future work the extensive evaluation of their proposed system towards swarming events and the extension of their experimentation towards other deep-learning CNN algorithms. Furthermore, the authors set as future work energy consumption of the Version 1 and 2 devices.

Author Contributions: Conceptualization, S.K.; methodology, S.K.; software, G.V.; validation, G.V. and S.K.; formal analysis, S.K. and G.V.; investigation, G.V. and S.K.; resources, S.K.; data curation, G.V.; writing—original draft preparation, G.V.; writing—review and editing, S.K.; visualization, G.V.; supervision, C.P.; project administration, S.K. All authors have read and agreed to the published version of the manuscript.

Funding: This research received no external funding.

Institutional Review Board Statement: Not applicable.

Informed Consent Statement: Not applicable.

Data Availability Statement: Not applicable.

Conflicts of Interest: Authors declare no conflict of interest.

Appendix A. Detection of Images Using the Faster-RCNN and SSD Algorithms with the Jetson Nano

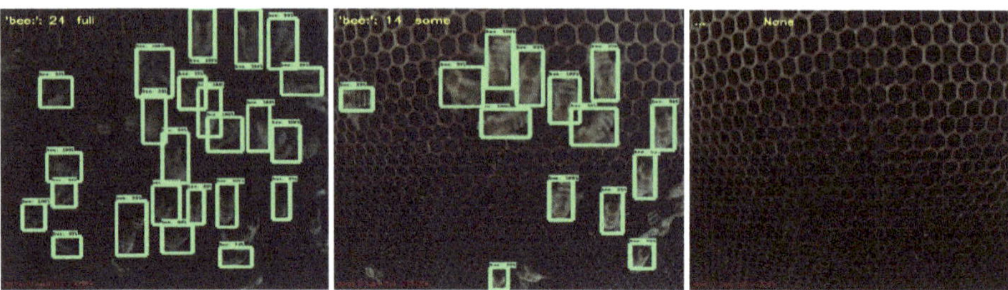

Figure A1. Bee detection with the NVIDIA Jetson Nano using the Faster-RCNN algorithm, Inception v2 model, for Classes 0, 2, and 3.

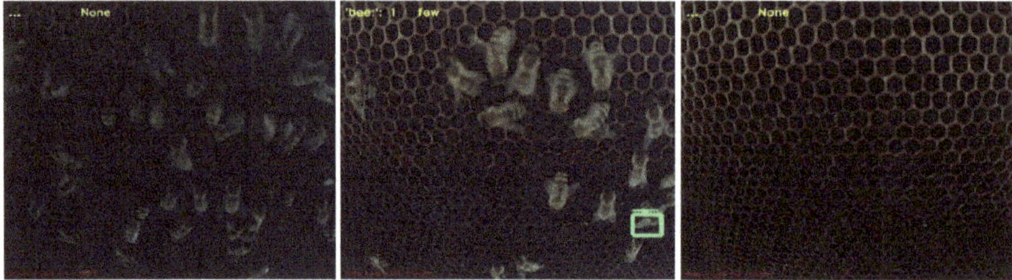

Figure A2. Bee detection with the NVIDIA Jetson Nano using the SSD algorithm, Inception v2 model, for Classes 0, 2, and 3.

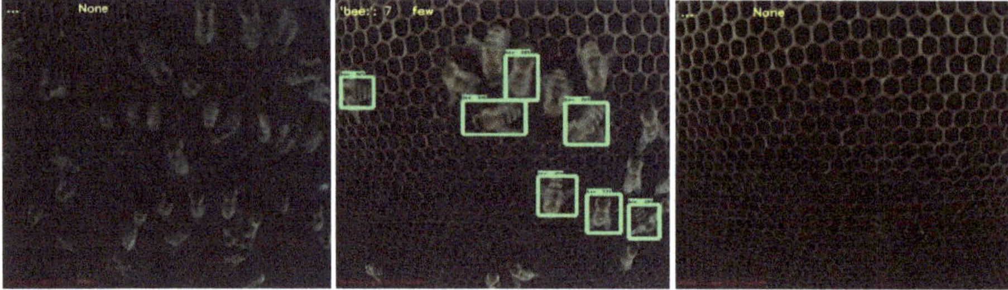

Figure A3. Bee detection with the NVIDIA Jetson Nano using the SSD algorithm, MobileNet v1 model, for Classes 0, 2, and 3.

Appendix B. BeeQ RMS Software

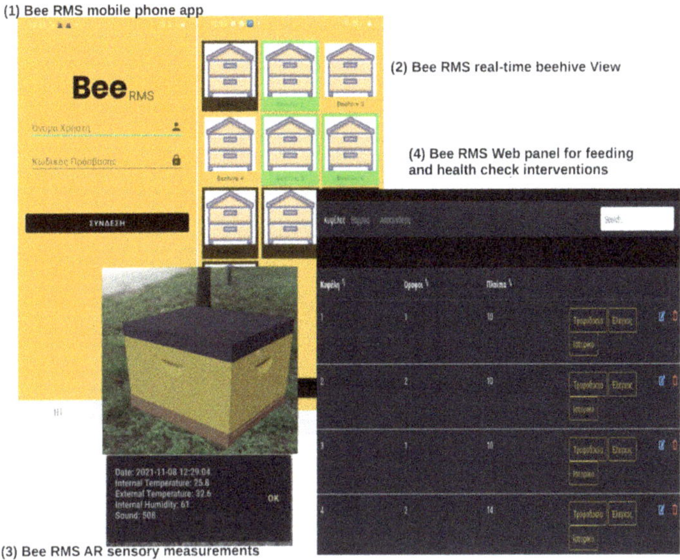

Figure A4. (1) BeeQ RMS system, (2) Android mobile phone application and sensory grid view of active beehives, (3) BeeQ RMS mobile phone sensory real-time measurements and AR beehive pose, and (4) BeeQ RMS web panel for feeding and health status check practices.

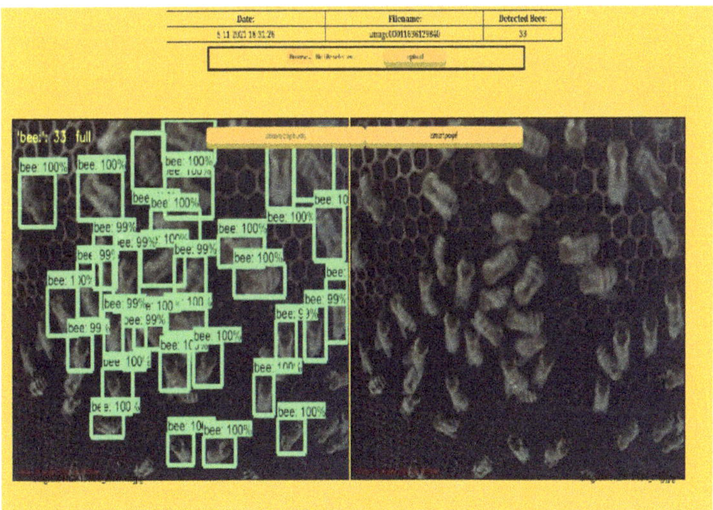

Figure A5. BeeQ RMS web interface to the swarming detection service. It takes as the input uploaded images and performs the offline detection process, illustrating the results.

Appendix C. BeeQ RMS End Node Devices (Versions 1 and 2)

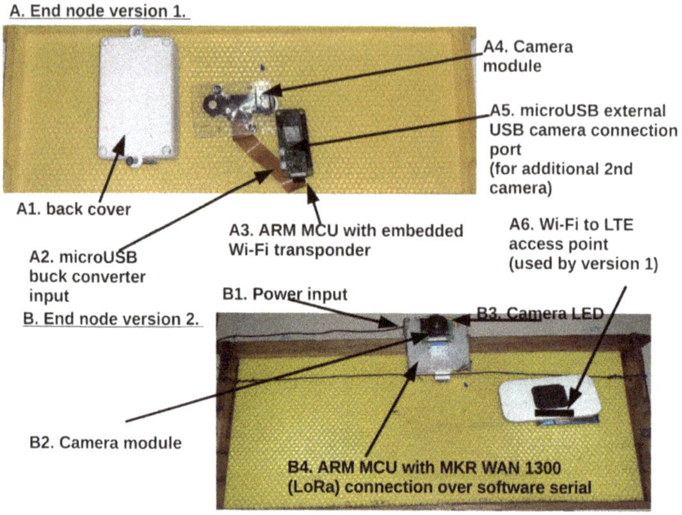

A. End node version 1.
A4. Camera module
A5. microUSB external USB camera connection port (for additional 2nd camera)
A1. back cover
A3. ARM MCU with embedded Wi-Fi transponder
A6. Wi-Fi to LTE access point (used by version 1)
A2. microUSB buck converter input
B1. Power input
B. End node version 2.
B3. Camera LED
B2. Camera module
B4. ARM MCU with MKR WAN 1300 (LoRa) connection over software serial

Figure A6. BeeQ RMS end node device prototypes (Version 1 using WiFi and LTE for images' data transmissions to the BeeQ RMS swarming detection service and the Version 2 device, which transmits the detection results over LoRa).

References

1. Abbasi, M.; Yaghmaee, M.H.; Rahnama, F. Internet of Things in agriculture: A survey. In Proceedings of the 2019 3rd International Conference on Internet of Things and Applications (IoT), Heraklion, Greece, 2–4 May 2019; Volume 1, pp. 1–12. [CrossRef]
2. Farooq, M.S.; Riaz, S.; Abid, A.; Abid, K.; Naeem, M.A. A Survey on the Role of IoT in Agriculture for the Implementation of Smart Farming. *IEEE Access* **2019**, *7*, 156237–156271. [CrossRef]
3. Farooq, M.S.; Riaz, S.; Abid, A.; Umer, T.; Zikria, Y.B. Role of IoT Technology in Agriculture: A Systematic Literature Review. *Electronics* **2020**, *9*, 319. [CrossRef]
4. Zinas, N.; Kontogiannis, S.; Kokkonis, G.; Valsamidis, S.; Kazanidis, I. Proposed Open Source Architecture for Long Range Monitoring. The Case Study of Cattle Tracking at Pogoniani. In *Proceedings of the 21st Pan-Hellenic Conference on Informatics*; ACM: New York, NY, USA, 2017. [CrossRef]
5. Kontogiannis, S. An Internet of Things-Based Low-Power Integrated Beekeeping Safety and Conditions Monitoring System. *Inventions* **2019**, *4*, 52. [CrossRef]
6. Mekala, M.S.; Viswanathan, P. A Survey: Smart agriculture IoT with cloud computing. In Proceedings of the 2017 International conference on Microelectronic Devices, Circuits and Systems (ICMDCS), Vellore, India, 10–12 August 2017; Volume 1, pp. 1–7. [CrossRef]
7. Zabasta, A.; Kunicina, N.; Kondratjevs, K.; Ribickis, L. IoT Approach Application for Development of Autonomous Beekeeping System. In Proceedings of the 2019 International Conference in Engineering Applications (ICEA), Azores, Portugal, 8–11 July 2019; Volume 1, pp. 1–6. [CrossRef]
8. Flores, J.M.; Gamiz, V.; Jimenez-Marin, A.; Flores-Cortes, A.; Gil-Lebrero, S.; Garrido, J.J.; Hernando, M.D. Impact of Varroa destructor and associated pathologies on the colony collapse disorder affecting honey bees. *Res. Vet. Sci.* **2021**, *135*, 85–95. [CrossRef] [PubMed]
9. Dineva, K.; Atanasova, T. ICT-Based Beekeeping Using IoT and Machine Learning. In *Distributed Computer and Communication Networks*; Springer International Publishing: Cham, Switzerland, 2018; pp. 132–143.
10. Olate-Olave, V.R.; Verde, M.; Vallejos, L.; Perez Raymonda, L.; Cortese, M.C.; Doorn, M. Bee Health and Productivity in Apis mellifera, a Consequence of Multiple Factors. *Vet. Sci.* **2021**, *8*, 76. [CrossRef] [PubMed]
11. Braga, A.R.; Gomes, D.G.; Rogers, R.; Hassler, E.E.; Freitas, B.M.; Cazier, J.A. A method for mining combined data from in-hive sensors, weather and apiary inspections to forecast the health status of honey bee colonies. *Comput. Electr. Agric.* **2020**, *169*, 105161. [CrossRef]
12. Jarimi, H.; Tapia-Brito, E.; Riffat, S. A Review on Thermoregulation Techniques in Honey Bees' (Apis Mellifera) Beehive Microclimate and Its Similarities to the Heating and Cooling Management in Buildings. *Future Cities Environ.* **2020**, *6*, 7. [CrossRef]

13. Peters, J.M.; Peleg, O.; Mahadevan, L. Collective ventilation in honeybee nests. *Future Cities Environ.* **2019**, *16*, 20180561. [CrossRef]
14. Zacepins, A.; Kviesis, A.; Stalidzans, E.; Liepniece, M.; Meitalovs, J. Remote detection of the swarming of honey bee colonies by single-point temperature monitoring. *Biosyst. Eng.* **2016**, *148*, 76–80. [CrossRef]
15. Zacepins, A.; Kviesis, A.; Pecka, A.; Osadcuks, V. Development of Internet of Things concept for Precision Beekeeping. In Proceedings of the 18th International Carpathian Control Conference (ICCC), Sinaia, Romania, 28–31 May 2017; Volume 1, pp. 23–27. [CrossRef]
16. Gil-Lebrero, S.; Quiles-Latorre, F.J.; Ortiz-Lopez, M.; Sanchez-Ruiz, V.; Gamiz-Lopez, V.; Luna-Rodriguez, J.J. Honey Bee Colonies Remote Monitoring System. *Sensors* **2017**, *17*, 55. [CrossRef]
17. Bellos, C.V.; Fyraridis, A.; Stergios, G.S.; Stefanou, K.A.; Kontogiannis, S. A Quality and disease control system for beekeeping. In Proceedings of the 2021 6th South-East Europe Design Automation, Computer Engineering, Computer Networks and Social Media Conference (SEEDA-CECNSM), Preveza, Greece, 24–26 September 2021; pp. 1–4. [CrossRef]
18. Ntawuzumunsi, E.; Kumaran, S.; Sibomana, L. Self-Powered Smart Beehive Monitoring and Control System (SBMaCS). *Sensors* **2021**, *21*, 3522. [CrossRef]
19. Prost, P.J.; Medori, P. *Apiculture*, 6th ed.; Intercept Ltd.: Oxford, UK, 1994.
20. Nikolaidis, I.N. *Beekeeping Modern Methods of Intensive Exploitation*, 11th ed.; Stamoulis Publications: Athens, Greece, 2005. (In Greek)
21. Clement, H. *Le Traite Rustica de L'apiculture*; Psichalos Publications: Athens, Greece, 2017.
22. He, W.; Zhang, S.; Hu, Z.; Zhang, J.; Liu, X.; Yu, C.; Yu, H. Field experimental study on a novel beehive integrated with solar thermal/photovoltaic system. *Sol. Energy* **2020**, *201*, 682–692. [CrossRef]
23. Kady, C.; Chedid, A.M.; Kortbawi, I.; Yaacoub, C.; Akl, A.; Daclin, N.; Trousset, F.; Pfister, F.; Zacharewicz, G. IoT-Driven Workflows for Risk Management and Control of Beehives. *Diversity* **2021**, *13*, 296. [CrossRef]
24. Terenzi, A.; Cecchi, S.; Spinsante, S. On the Importance of the Sound Emitted by Honey Bee Hives. *Vet. Sci.* **2020**, *7*, 168. [CrossRef] [PubMed]
25. Ferrari, S.; Silva, M.; Guarino, M.; Berckmans, D. Monitoring of swarming sounds in beehives for early detection of the swarming period. *Comput. Electron. Agric.* **2008**, *64*, 72–77. [CrossRef]
26. Nolasco, I.; Terenzi, A.; Cecchi, S.; Orcioni, S.; Bear, H.L.; Benetos, E. Audio-based Identification of Beehive States. In Proceedings of the IEEE International Conference on Acoustics, Speech and Signal Processing (ICASSP), Brighton, UK, 12–17 May 2019; Volume 1, pp. 8256–8260. [CrossRef]
27. Zgank, A. Bee Swarm Activity Acoustic Classification for an IoT-Based Farm Service. *Sensors* **2020**, *20*, 21. [CrossRef]
28. Nolasco, I.; Benetos, E. To bee or not to bee: Investigating machine learning approaches for beehive sound recognition. *CoRR* **2018**. Available online: http://xxx.lanl.gov/abs/1811.06016 (accessed on 14 April 2019).
29. Liew, L.H.; Lee, B.Y.; Chan, M. Cell detection for bee comb images using Circular Hough Transformation. In Proceedings of the 2010 International Conference on Science and Social Research (CSSR 2010), Kuala Lumpur, Malaysia, 5–7 December 2010; pp. 191–195. [CrossRef]
30. Baptiste, M.; Ekszterowicz, G.; Laurent, J.; Rival, M.; Pfister, F. Bee Hive Traffic Monitoring by Tracking Bee Flight Paths. 2018. Available online: https://hal.archives-ouvertes.fr/hal-01940300/document (accessed on 5 September 2019).
31. Simic, M.; Starcevic, V.; Kezić, N.; Babic, Z. Simple and Low-Cost Electronic System for Honey Bee Counting. In Proceedings of the 28th International Electrotechnical and Computer Science Conference, Ambato, Ecuador, 23–24 September 2019.
32. Bee-Shop Security Systems: Surveillance Camera for Bees. 2015. Available online: http://www.bee-shop.gr (accessed on 14 June 2020).
33. EyeSon Hives Honey Bee Health Monitor. | Keltronix. 2018. Available online: https://www.keltronixinc.com/ (accessed on 7 June 2018).
34. Theodoros Belogiannis. Zygi Beekeeping Scales with Monitoring Camera Module. 2018. Available online: https://zygi.gr/en (accessed on 10 March 2021).
35. Arnia Remote Hive Monitoring System. Better Knowledge for Bee Health. 2017. Available online: https://arnia.co.uk (accessed on 4 June 2018).
36. Hive-Tech 2 Crowd Monitoring System for Your Hives. 2019. Available online: https://www.3bee.com/en/crowd/ (accessed on 16 March 2019).
37. Hivemind System to Monitor Your Hives to Improve Honey Production. 2017. Available online: https://hivemind.nz/for/honey/ (accessed on 10 March 2021).
38. Hudson, T. Easy Bee Counter. 2018. Available online: https://www.instructables.com/Easy-Bee-Counter/ (accessed on 8 September 2020).
39. Hudson, T. Honey Bee Counter II. 2020. Available online: https://www.instructables.com/Honey-Bee-Counter-II/ (accessed on 8 September 2020).
40. Gomez, K.; Riggio, R.; Rasheed, T.; Granelli, F. Analysing the energy consumption behaviour of WiFi networks. In Proceedings of the 2011 IEEE Online Conference on Green Communications, Online Conference, Piscataway, NJ, USA, 26–29 September 2011; Volume 1, pp. 98–104. [CrossRef]

41. Nurgaliyev, M.; Saymbetov, A.; Yashchyshyn, Y.; Kuttybay, N.; Tukymbekov, D. Prediction of energy consumption for LoRa based wireless sensors network. *Wirel. Netw.* **2020**, *26*, 3507–3520. [CrossRef]
42. Lavric, A.; Valentin, P. Performance Evaluation of LoRaWAN Communication Scalability in Large-Scale Wireless Sensor Networks. *Wirel. Commun. Mob. Comput.* **2018**, *2018*, 6730719. [CrossRef]
43. Van den Abeele, F.; Haxhibeqiri, J.; Moerman, I.; Hoebeke, J. Scalability Analysis of Large-Scale LoRaWAN Networks in ns-3. *IEEE Internet Things J.* **2017**, *4*, 2186–2198. [CrossRef]
44. Haxhibeqiri, J.; De Poorter, E.; Moerman, I.; Hoebeke, J. A Survey of LoRaWAN for IoT: From Technology to Application. *Sensors* **2018**, *18*, 3995. [CrossRef]
45. Mokar, M.A.; Fageeri, S.O.; Fattoh, S.E. Using Firebase Cloud Messaging to Control Mobile Applications. In Proceedings of the International Conference on Computer, Control, Electrical, and Electronics Engineering (ICCCEEE), Khartoum, Sudan, 21–23 September 2019; pp. 1–5. [CrossRef]
46. Wang, W.; Yang, Y.; Wang, X.; Wang, W.; Li, J. Development of convolutional neural network and its application in image classification: A survey. *Opt. Eng.* **2019**, *58*, 1–19. [CrossRef]
47. Bharati, P.; Pramanik, A. Deep Learning Techniques-R-CNN to Mask R-CNN: A Survey. In *Computational Intelligence in Pattern Recognition*; Springer: Singapore, 2020; pp. 657–668.
48. Tzudalin, D. LabelImg Is a Graphical Image Annotation Tool and Label Object Bounding Boxes in Images. 2016. Available online: https://github.com/tzutalin/labelImg (accessed on 20 September 2019).
49. Labelbox. Labelbox: The Leading Training Data Platform for Data Labeling. Available online: https://labelbox.com (accessed on 2 June 2021).
50. Image Annotation Tool. Available online: https://github.com/alexklaeser/imgAnnotation (accessed on 2 June 2021).
51. Computer Vision Annotation Tool (CVAT). 2021. Available online: https://github.com/openvinotoolkit/cvat (accessed on 2 June 2021).
52. Lim, B.; Son, S.; Kim, H.; Nah, S.; Lee, K.M. Enhanced Deep Residual Networks for Single Image Super-Resolution. In Proceedings of the IEEE Conference on Computer Vision and Pattern Recognition (CVPR) Workshops, Honolulu, HI, USA, 21–26 July 2017; Volume 1, pp. 1132–1140. [CrossRef]
53. Bradski, G. The OpenCV Library. *Dr. Dobb's J. Softw. Tools* **2000**, *25*, 120–123.
54. GitHub-Tensorflow/Models: Models and Examples Built with TensorFlow 1. Available online: https://github.com/tensorflow/models/tree/r1.12.0 (accessed on 15 September 2018).
55. GitHub-Tensorflow/Models: Models and Examples Built with TensorFlow 2. 2021. Available online: https://github.com/tensorflow/models (accessed on 12 November 2020).
56. Abadi, M.; Agarwal, A.; Barham, P.; Brevdo, E.; Chen, Z.; Citro, C.; Corrado, G.S.; Davis, A.; Dean, J.; Devin, M.; et al. TensorFlow: Large-Scale Machine Learning on Heterogeneous Systems. 2015. Available online: https://www.tensorflow.org/ (accessed on 20 September 2018).
57. TensorFlow GPU Support. 2017. Available online: https://www.tensorflow.org/install/gpu?hl=el (accessed on 15 September 2018).
58. Moroney, L. The Firebase Realtime Database. In *The Definite Guide to Firebase*; Apress: Berkeley, CA, USA, 2017; pp. 51–71. [CrossRef]
59. Huang, J.; Rathod, V.; Sun, C.; Zhu, M.; Korattikara, A.; Fathi, A.; Fischer, I.; Wojna, Z.; Song, Y.; Guadarrama, S.; et al. Speed and accuracy trade-offs for modern convolutional object detectors. *arXiv* **2017**, arXiv:1611.10012.

Article

Radio Wave Attenuation Measurement System Based on RSSI for Precision Agriculture: Application to Tomato Greenhouses

Dora Cama-Pinto [1,*], Juan Antonio Holgado-Terriza [2], Miguel Damas-Hermoso [1], Francisco Gómez-Mula [1] and Alejandro Cama-Pinto [3,*]

1 Department of Computer Architecture and Technology, University of Granada, 18071 Granada, Spain; mdamas@ugr.es (M.D.-H.); frgomez@ugr.es (F.G.-M.)
2 Department of Software Engineering, University of Granada, 18071 Granada, Spain; jholgado@ugr.es
3 Department of Computer Science and Electronics, Universidad de la Costa, Barranquilla 080002, Colombia
* Correspondence: doracamapinto@correo.ugr.es (D.C.-P.); acama1@cuc.edu.co (A.C.-P.); Tel.: +57-5-3225498 (A.C.-P.)

Abstract: Precision agriculture and smart farming are concepts that are acquiring an important boom due to their relationship with the Internet of Things (IoT), especially in the search for new mechanisms and procedures that allow for sustainable and efficient agriculture to meet future demand from an increasing population. Both concepts require the deployment of sensor networks that monitor agricultural variables for the integration of spatial and temporal agricultural data. This paper presents a system that has been developed to measure the attenuation of radio waves in the 2.4 GHz free band (ISM- Industrial, Scientific and Medical) when propagating inside a tomato greenhouse based on the received signal strength indicator (RSSI), and a procedure for using the system to measure RSSI at different distances and heights. The system is based on Zolertia Re-Mote nodes with the Contiki operating system and a Raspberry Pi to record the data obtained. The receiver node records the RSSI at different locations in the greenhouse with the transmitter node and at different heights. In addition, a study of the radio wave attenuation was measured in a tomato greenhouse, and we publish the corresponding obtained dataset in order to share with the research community.

Keywords: wireless sensor networks; WSN; received signal strength indicator; RSSI; Internet of Things; IoT; free space pathloss; smart farming

check for updates

Citation: Cama-Pinto, D.; Holgado-Terriza, J.A.; Damas-Hermoso, M.; Gómez-Mula, F.; Cama-Pinto, A. Radio Wave Attenuation Measurement System Based on RSSI for Precision Agriculture: Application to Tomato Greenhouses. *Inventions* **2021**, *6*, 66. https://doi.org/10.3390/inventions6040066

Academic Editor: Konstantinos G. Arvanitis

Received: 8 September 2021
Accepted: 8 October 2021
Published: 12 October 2021

1. Introduction

Agriculture is a fundamental pillar in a country's economy and society; therefore, increasing agricultural productivity through cutting-edge technology contributes to economic progress and the feeding of its inhabitants [1]. In this perspective, technological advancement in precision agriculture (PA) has been accelerated by the Internet of Things (IoT), wireless sensor networks (WSN), and even 5G networks [2,3], because almost all techniques used in PA have sensing technology as a common factor [4,5]. PA produces more food with limited resources such as water and soil [6–8], effectively increasing the quality and yield of crops [9], a remarkable fact considering the Food and Agriculture Organization of the United Nations (FAO) states that the human population level on a global scale will increase from eight billion inhabitants in 2025 to nine billion, six hundred million in 2050 [10].

In a PA scenario, wireless sensor networks (WSNs) serve as a local crop monitoring system that allows for making the right decisions in a controlled production system affected by climate change [11–14]. Sensor values in agricultural fields are adjusted and set according to the specific requirements of each type of plant, e.g., in an appropriate range to provide continuous information on the field conditions of the temperature, wind, light, soil moisture, nutrients, and illuminance variables [15–17].

On the other hand, with WSN being a distributed system, it is usually composed of small-sized embedded devices [18] called nodes or motes, which communicate wirelessly in a network. Each node has four subsystems described in Figure 1. The first detects the environment by means of its sensors that measure physical data and translate them into analogue signals that are then converted into digital signals. The second is the processing subsystem which contains the microcontroller that performs calculations on the digital data collected through the ADC. The third is the communication subsystem that is in charge of exchanging the information between the different sensor nodes by means of its transceivers [3]. Radio modules operating according to the IEEE 802.15.14 standard typically use radio frequency transceivers in the 2.4 GHz band (2400–2483.5 MHz) [19], as it is freely available worldwide and the most widely used among radio modules from manufacturers using WLAN [20–22], PAN wireless communication standards such as IEEE 802.15.1 (Bluetooth), IEEE 802.15.4 used in WSN (wireless sensor network), and IEEE 802.11 (WiFi). Finally, the fourth subsystem is the power subsystem, with a design based on energy saving, which is relevant as sensors have to operate on an extremely limited energy budget [10,23].

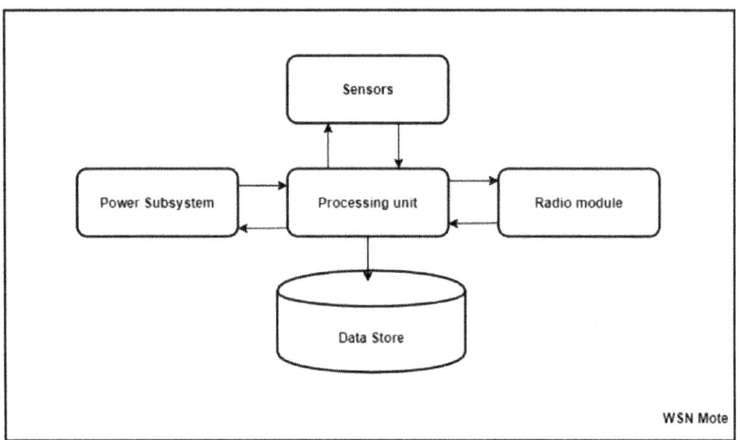

Figure 1. Schematic of the four subsystems of each node.

Wireless data transmission is one of the main features of IoT which enables information collection on a wide spectrum of physical parameters in agriculture thanks to its self-organization, low cost, low power consumption, wide coverage area, and deployment in complex and changing environments over areas with problematic power supply among its other characteristics [24]. This facilitates making farming practices more productive, sustainable and environmentally friendly [25–27].

In this work we are interested in the development of a new measurement system that facilitates the registration of the RSSI (received signal strength indicator) of radio waves in the 2.4 GHz band that propagate in greenhouse crops for carrying out specific studies of radio wave attenuation to validate the developed system, in our case for tomato greenhouses. Other works have developed a similar RSSI measurement system design but specifically to find the positioning of nodes inside a greenhouse [28–30]. However, our measurement system is prepared especially for measuring the radio wave attenuation inside greenhouses.

Some parts of this work were presented previously in [31]. However, this work includes not only the novel RSSI measurement system, but also the procedure to be used for measuring the radio wave attenuation in greenhouses.

In next sections, we present how the radio wave attenuation measurement system based on RSSI was developed, detailing its configuration, assembly and deployment on greenhouses. The article is divided into six sections. Section 2 describes the architecture, hardware, and software of the developed measurement and recording system. Section 3 details how the system was deployed during field tests and the steps followed to perform the measurements. Section 4 describes details of the registered dataset obtained in specific tomato greenhouses. Section 5 discusses the results obtained, and Section 6 presents the conclusions of our work.

2. Radio Wave Attenuation Measurement System in a Greenhouse

Part of the agricultural industry is based on greenhouse production systems, originally implemented in northern latitudes or geographic areas with a cold climate, so that the harvesting season would be prolonged over the year. In that sense, because the greenhouse plantations are located in well-defined indoor areas, WSNs are easier to implement than in outdoor crops in the field. The amount of transmitted data for WSNs used in agriculture is usually small, as the record of each monitored sample is taken at a considerably spaced time because its values change at a slow rate. Still, constant data transmission is necessary while minimizing the number of IoT devices deployed in the field, in order to reduce the cost of implementation and deployment of the nodes in the crop area, while ensuring the performance of the system [32]. One way to reduce the number of nodes in the WSN is to deploy them at the maximum distance from each other. However, the maximum spacing of the nodes varies in relation to the type of crop monitored [33]. Therefore, it is taken into account that crop growth can affect the propagation of electromagnetic signals, the deployment of WSN nodes and topology control [12], because the wireless signal, in addition to losing power with increasing distance between nodes, suffers from multipath fading due to reflection, diffraction and scattering when the radio signal propagates through the environment between the transmitter and receiver due to the presence of the leafy branches, leaves, and fruits of crops [9,34]. At this point, there is a clear interest of the scientific community in propagation studies and planning in the deployment of WSNs using different radio wave frequencies in different densities of food crops, and developing propagation models to establish the loss in the signal path [35–37].

In the case of our developed measurement system, this allows measurements of the received signal strength level (RSSI) inside a farm, and in particular in a greenhouse at different distances and height between the transmitter and the receiver. Also, this system contributes to precision agriculture by determining the maximum communication distance between two wireless nodes, to efficiently plan the number of nodes in the WSN and their coverage area in sensor/actuator deployment within an agricultural field. There are other studies on this subject, which also employ wireless sensor networks, and use mobile devices based on Arduino boards and Xbee radio modules [38–40]. In our case, for the RSSI measurement of radio waves, we use a board that has an integrated radio module in the 2.4 GHz ISM band [41]. On the other hand, we include data logging in the system itself so that it is autonomous, and can operate 24 h for two weeks without interruption. The main benefit of the developed system is its portability, ease of installation in an agricultural environment, and sufficient autonomy time for continuous monitoring. The system is also capable of operating in the 868/915 MHz band, although no tests were carried out on this band in this study. If the working environment inside a greenhouse is difficult for the farmer because it is a closed environment, usually with higher temperature and humidity levels than outside, it is even more difficult for someone who is not involved in these tasks, such as a researcher who performs radio wave measurements for hours inside a greenhouse. In this sense, the main difference of our proposed system concerning similar solutions is its ease of implementation, verified in our field tests, and its ability to store data that can then be used to evaluate the attenuation behavior of radio waves in a crop with the possibility of generating new models from the values taken. The greater the amount of data collected, the greater the accuracy of the results obtained in the analysis, so our system can record RSSI for a full day unattended.

2.1. Architecture

The architecture of the system is composed of two stations, the receiving station which records the RSSI of the signal sent from the transmitting station. The receiving station is composed of the Re-Mote node which is connected and powered through its USB port connected to a Raspberry Pi embedded system, and the latter to a 220 V power socket inside the greenhouse. The system is autonomous, but for monitoring purposes, it is connected sporadically to a laptop with a UTP cable to the RJ45 port of the Raspberry Pi. The transmitter station is made up of the Re-Mote node, which is powered by a rechargeable 3.7 V lithium-ion battery with a nominal capacity of 6600 mAh that gives it autonomy in its operation. These elements are housed in a PVC enclosure with an IP65 protection rating, prepared to keep out dust and water jets.

Figure 2 shows schematically the arrangement of the two stations in the crop. Each station is placed on a mast supported by a 17 kg base for stability. Acrylic plates that carry the PVC boxes are attached to the masts. The PVC box at the transmitter station houses the Re-Mote node and a battery that powers it, giving it autonomy. On the receiver side, the PVC box houses the Re-Mote node intercommunicated and powered by its USB connection to the Raspberry Pi, which is the data logging unit, with its charger plugged into a 220 V socket.

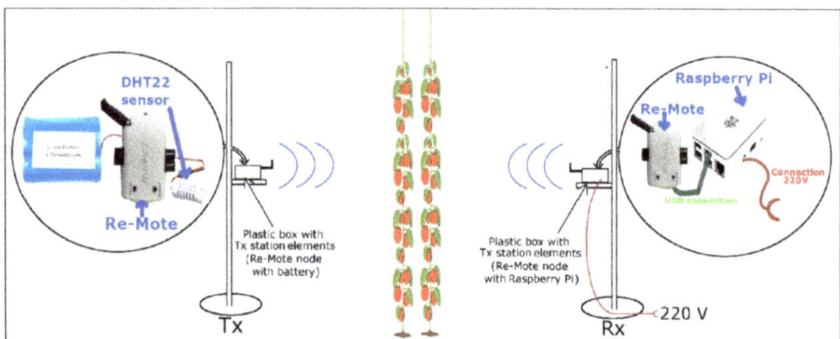

Figure 2. Wireless communication scheme of Tx and Rx stations in a tomato greenhouse. Tx station (**left**) powered by Lithium-ion 3.7 V-6600 mAh battery. Rx station (**right**) connected to a Raspberry Pi embedded computer.

2.2. System Hardware

The measuring and recording system consists of the following hardware components:

- Re-Mote nodes. Zolertia's Re-Mote motes have radio transceiver modules capable of acting as a Tx transmitter and Rx receiver node. It was chosen because its board is an IoT platform with extensive Contiki OS software support, including 6LoWPAN, RPL and other widely used IoT protocols. It integrates Texas Instruments' CC2538 System-on-Chip (SoC) chip for low-power, short-range communication in the 2.4 GHz band, with a current consumption of 24 mA when transmitting, 20 mA when receiving, and 1.3 uA in the sleep state [42–44]. On the other hand, the EIRP (equivalent radiated isotropic radiated power) of the Tx node was −29 dBm in the tests and 5 dBi gain antennas were used for both motes. The re-receive sensitivity at the Re-Mote nodes was −97 dBm.
- Raspberry Pi. A Raspberry Pi was selected as a server that stores the measurement data on an SD memory stick in CSV format (Figure 2). It is co-connected and powered to the Rx-mote through its USB port [45].
- Lithium-ion battery. The lithium-ion battery was connected to the 3.7 V Tx transmitter to maintain the transmitter's autonomy.
- Humidity and temperature sensor. In addition, the DHT22 sensor was connected to the transmitter module to transmit temperature and humidity data, which are

traditionally used to monitor and supervise the environmental conditions of the crop in a greenhouse.

2.3. System Software

For the proper functioning of the system, it is necessary to develop the appropriate software infrastructure in order to minimize energy consumption. The software developed in each case is specified below.

- Re-Mote nodes. The Contiki operating system, developed in 2002 by Adam Dunkels, was installed and configured as an open source runtime environment for low-power, memory-limited wireless sensor nodes [46]. It is lightweight, making it ideal for IoT. Its applications are developed with the C programming language. It has a built-in TCP/IP implementation for embedded devices, officially supporting various device platforms that make up wireless sensor networks, including the Re-Mote board [47–49]. Only Contiki's power-saving module (power-mgmt.h) was used in the transmitter node, because during the test phase this station is the one that is far away from the receiver node and does not have an electrical outlet, instead being powered by its own battery. The receiving node was powered by the Raspberry Pi, which in turn was connected to a power socket at one end of the greenhouse. For radio communication, we employed the Rime stack (rime.h), which provides a set of basic communication primitives for best-effort single-hop network broadcasting ("unicast") and reliable multi-hop "multi-hop unicast" [50].
- The program developed in C for the transmitter station sends frames of temperature and humidity data obtained from the DHT22 sensor periodically on a variable timer, while remaining suspended the rest of the time. This reduces power consumption, extending the transmitter's autonomy. On the other hand, the receiving station, powered by the Raspberry Pi and the greenhouse socket, measures the RSSI and obtains the data frame sent by the transmitter node.
- Raspberry Pi. The Raspbian distribution based on Debian was installed and several scripts were developed in the Python language that established serial communication with the Zolertia devices and generated.csv files with the data they receive, storing them in the SD memory of the Raspberry Pi. It also has a clock module with a CR2032 battery so that the date and time are not decalibrated when it is turned off, recording it with each RSSI record.

3. Deployment and Commissioning in a Tomato Greenhouse

This section details how the system was deployed, and how the experiments were designed in a tomato greenhouse located in the province of Almeria, Autonomous Community of Andalusia-Spain. This product, in addition to its nutritional value, is in high demand in the European Community markets.

3.1. Deployment of the System

The Tx and Rx (Figure 2) nodes were placed at different distances inside the greenhouse and at different heights with the help of masts. The station was placed on a 17 kg mast base for stability (Figure 3B). Both stations were installed at the same height, and the height was varied throughout the measurements. In general, the tomato plants were aligned in a wall and equidistant from each other, separated by a space or lane where the farmers walk, as seen in Figure 3A,C.

Figure 3. (**A**) Tomato crop side profile. (**B**) Location of transmitter node inside the greenhouse. (**C**) View of greenhouse passageway.

3.2. Conducting Experiments

For the experiments, values were recorded every 10 s for 10 min at each position. The height of the nodes was the same for the Tx and Rx node at each stage of the measurement. In addition to the RSSI recording and analysis tests, monitoring functions for agricultural and environmental variables can also be performed by plugging analogue and digital sensors, e.g., humidity or temperature, into the Re-Mote.

3.3. Measurement Procedure

With the proposed system, different types of studies can be carried out to identify radio wave propagation patterns in orchards [51], mango [52], and tomato greenhouses. In these studies, we should place the receiver and transmitter node at different distances and heights. The RSSI values recorded can be taken at different stages of the crop, from planting to harvesting if it is desired to analyze the data and/or build a model relating radio wave attenuation to the growth stages of the plantation. Alternatively, the analysis or model generation can be developed from the recorded data, when the data are taken in the extreme case, i.e., near harvest time, when the canopy thickness is at its maximum.

In our case we follow the following steps to carry out a study of radio wave attenuation based on RSSI, and is summarized in Figure 4:

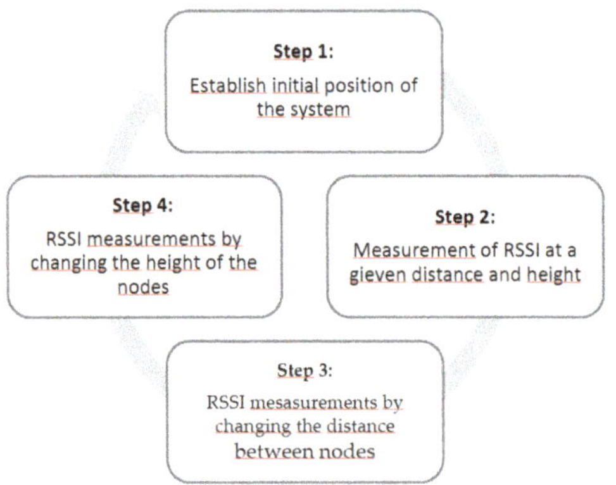

Figure 4. Steps to carry out a radio attenuation study based on RSSI inside a greenhouse.

Step 1: Initial position of the system. The position of the receiving station is fixed, as it is powered by a 220 V electrical current from a socket in the greenhouse. In any study with the system, the receiving station depicted as Rx is the reference node where the RSSI is to be measured. For instance, in Figure 5A the positions of the receiving nodes are shown in red (A1, A2, A3, A4) at one end of the greenhouse during the measurement of RSSI.

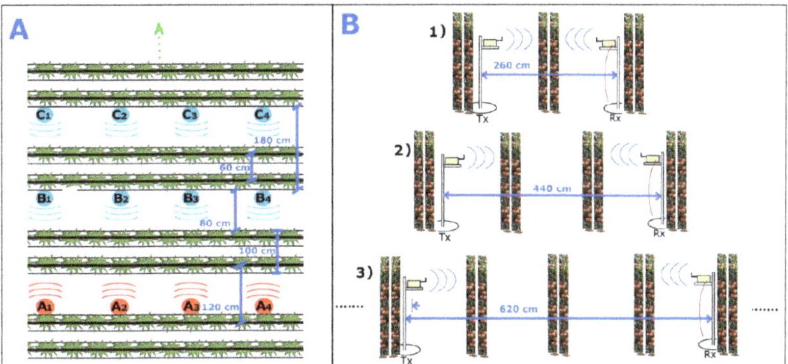

Figure 5. (**A**) The figure on the left shows a planting frame inside the greenhouse where the transmitting node (light blue) changes the distance with respect to the receiving node (red). (**B**) On the right the figure shows that the height of both nodes was maintained constant while the distance between them changed.

Step 2: RSSI Measurement at a specific distance and height. In this case, the transmitting station depicted as Tx is placed at a specific position with respect to the Rx node. According to the Figure 5A, the RX node was fixed initially at the position A1 while the Tx node was placed at position B1 with a specific distance of 260 cm. This distance depends on the row of plants and the spacing between the rails in the greenhouse. In this case, there is only one row of spacing plants between them. In addition to the distance, both nodes must also be situated at the same height, as shown in Figure 5B.

Once the distance and height are fixed for both nodes, the receiver node can perform the RSSI measurements and collect at the same time the temperature and humidity measured by transmitter node to be transmitted later to receiver node. These measurements can be performed at a customizable rate and can be stored as a data frame.

Subsequently, both stations are moved to the right (for instance, A2 and B2 in Figure 5) maintaining the same height and distance between them. New RSSI measurements were performed by Rx node at a constant rate, generating a new data frame. The process (node movement and measurement) was reiterated until the locations A4 and B4 were reached. The repeated measurements at the same distance and height in different positions guarantee that the measurement of RSII values take into account the fluctuations that occurred by the presence of the leafy branches, leaves, and fruits of crops.

A data collection of all these measurements at different positions (A1, A2, A3, A4) is then stored in the Raspberry PI of the Rx node.

Step 3: RSSI Measurements: changing the distance. In this case, the two stations are moved away from each other, changing the distance between them. For instance, in Figure 5B we can see that after the RSSI measurement in position B, we can move the Tx station to position C 440 cm away, which corresponds to the distance between two rows of plants between the transmitting and receiving station. The Tx station was placed at positions C1, C2, C3, C4, and the RSSI was measured and recorded with the Rx station at positions A1, A2, A3, A4 respectively. On the other hand, we can see in Figure 5B that the height was the same during the RSSI measurements.

Step 4: RSSI Measurements: changing the height. In this case the two stations changed their heights to a new value at the same time, and then the measurement procedure at different distances was repeated again. After the Tx station reaches the opposite end of the greenhouse, the procedure is repeated by changing the height of both nodes to a new height, e.g., 50 cm above the ground. The steps are then repeated starting from step 1.

4. Radio Wave Attenuation Dataset in Tomato Greenhouse

In order to validate the measurement system as well as the measurement procedure, a study was performed of radio wave attenuation based on RSSI in tomato greenhouses, specifically in a greenhouse at the Cañada de San Urbano in Almeria (Spain). It is an intensive farming area at the south of Spain, with many greenhouses providing fruit and vegetables to Europe. Figure 6A shows the location on a map of Spain, and Figure 6B shows a satellite image of the greenhouse in Almeria.

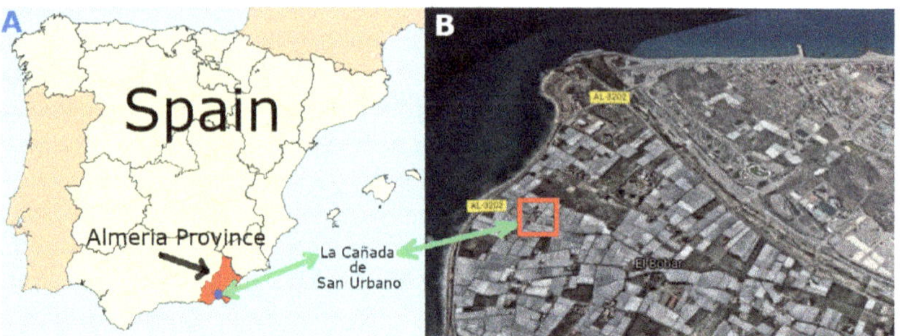

Figure 6. Location of the greenhouse in the Cañada de San Urbano in the province of Almeria. (**A**) On the map of Spain. (**B**) Satellite Image.

In this case, the RSSI measurements were obtained by the receiver node at different distances (260, 440, 620, 800, 980, 1160, 1340, 1520, 1700, 1880, 2060, 2240, and 2420 cm) with different heights (30, 50, 70, 90, 100, 150, and 200 cm). Initially, the receiver node placed at 5 cm from transmitter node reported an RSSI value of −24 dBm (best signal strength); this was the RSSI reference value used to determine the radio wave attenuation when the receiver was moved away from the transmitter node.

A recording of all measurements was saved on a dataset. The RSSI measurements at a specific position (for example A1) corresponding to a data frame were taken every 10 s and stored in CSV format inside the SD memory of the Raspberry Pi. The data frames of different positions at the same distance and height are stored in the same file. Each data frame included values of the RSSI in dBm, the ambient temperature and humidity, and the timestamp including date and time.

The file names indicate first the height of the node antennas, then the distance between the nodes. For example, the file "50–1880 csv" includes the record of the measurements with the antennas of the nodes at 50 cm and a distance of 1880 cm between them. In each file there are 4 segments corresponding to the change of positions of the nodes Tx and Rx, e.g., A1 with B1, A2 with B2, A3 with B3, and A4 with B4, as shown in Figure 5. This dataset can be found at the following link [53]. Table 1 is an example detailing the organization of the values. The first column indicates the sample number, then the temperature value in degrees Celsius, the relative humidity, the RSSI in dBm, the date, and the time the sample was collected.

Table 1. Steps to carry out a radio attenuation study based on RSSI inside a greenhouse.

Cycle	Temperature	Humidity	RSSI	Date	Hour
1	14	63	−93	25 March 2018	11:37:48
2	14	63	−93	25 March 2018	11:37:58
3	14	63	−93	25 March 2018	11:38:08
4	14	63	−93	25 March 2018	11:38:18
5	14	63	−93	25 March 2018	11:38:28

Table 1. *Cont.*

Cycle	Temperature	Humidity	RSSI	Date	Hour
6	14	63	−94	25 March 2018	11:38:38
7	14	63	−94	25 March 2018	11:38:48
8	14	63	−94	25 March 2018	11:38:58
9	14	63	−94	25 March 2018	11:39:08
10	14	63	−93	25 March 2018	11:39:18
11	14	63	−93	25 March 2018	11:39:28
12	14	63	−93	25 March 2018	11:39:38
13	14	63	−93	25 March 2018	11:39:48
14	14	63	−92	25 March 2018	11:39:58
15	14	63	−91	25 March 2018	11:40:08
16	14	63	−91	25 March 2018	11:40:18
17	14	63	−91	25 March 2018	11:40:28
18	14	63	−91	25 March 2018	11:40:38
19	14	63	−91	25 March 2018	11:40:48
20	14	63	−91	25 March 2018	11:40:58
21	14	63	−91	25 March 2018	11:41:08
22	14	63	−91	25 March 2018	11:41:18
23	14	63	−91	25 March 2018	11:41:28
24	14	63	−91	25 March 2018	11:41:38
25	14	63	−91	25 March 2018	11:41:48

In our measurement, the power-saving module "power-mgmt.h" was used on the Zolertia Re-Mote nodes, which caused the battery to go from 3.7 V (Tx node) to 2.8 V in about 60 h of use, as measurements were taken every 10 s. Below 2.8 V, it is advisable not to make measurements because the values recorded by the sensors may contain spurious errors.

5. Discussion

From the measurements carried out at different distances and heights, we can analyze the average value of the RSSI measurements and study how the radio wave signals at 2400 MHz attenuate as they pass through the rows of tomato plants (Figure 7). In Figure 7A, we observe that at 50 cm above the ground the maximum distance between the communication of the two Tx and Rx nodes is 2420 cm, and the minimum distance of coverage between them (1340 cm) is reached when the nodes are placed 150 cm above the ground. The values were recorded at the receiver node until they approached −100 dBm, because the receiver sensitivity is −97 dBm. However, for the purpose of wireless link stability, it is suggested to have a margin between the power detected by the receiver and the receiver sensitivity equal to or greater than 10 dB [54]. On the other hand, our system based on Re-Mote nodes is more compact than other, larger systems with a non-integrated transceiver, and the developed system has a unit price of approximately 130 euros, much lower than other options on the market.

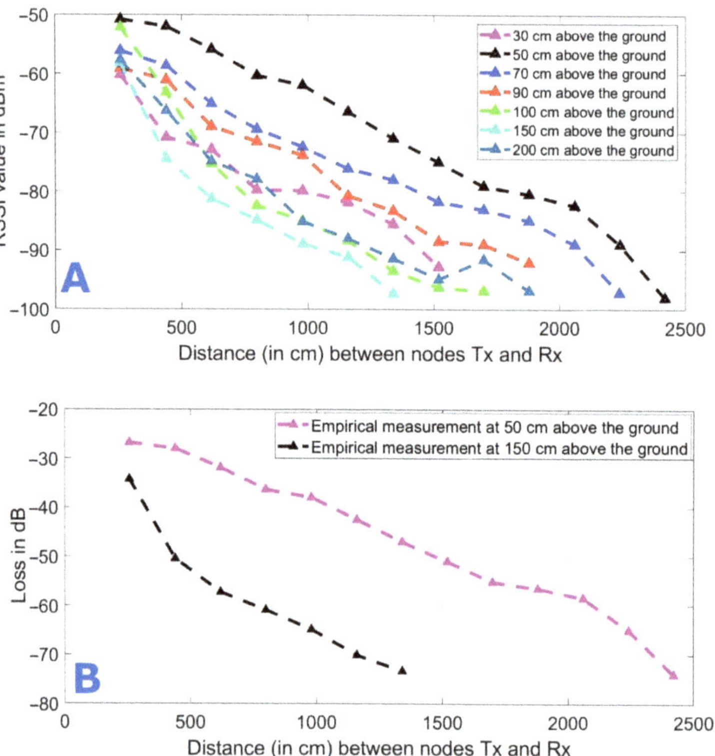

Figure 7. (**A**) Signal power levels in dBm between the transmitter (Tx) and receiver (Rx) nodes at different heights and distances between them. (**B**) Path loss or attenuation measured in dB of the radio wave between the Tx and Rx node inside the tomato greenhouse at 50 cm and 150 cm above the ground [41].

Figure 7B shows the radio wave attenuation curves between the Tx and Rx nodes at 50 and 150 cm from the ground. For example, the average RSSI value at 260 cm distance between the two nodes and 150 cm from the ground is −58.22 dBm (See Figure 7A). Thus, the path attenuation is obtained with the following calculation: EIRP + Path_attenuation + Gain_RX= −58.22 dBm.

Therefore,

$$\text{Path\_attenuation} = -58.22 \text{ dBm} + 29 \text{ dBm} - 5 \text{ dBi} \quad (1)$$

$$\text{Path\_attenuation} = -34.22 \text{ dB} \quad (2)$$

6. Conclusions and Future Work

The results obtained serve to improve the wireless coverage planning of radio waves operating in the 2.4 GHz frequency band inside a greenhouse in order to determine the best location and height of the Tx and Rx nodes with respect to the ground in a one-hop communication between them, as well as to minimize their number in the deployment.

The collected measurements establish that the maximum one-hop distance between Tx and Rx within a tomato greenhouse is recorded when the antenna is located 50 cm above the ground, while the lowest coverage occurs when the antenna nodes are 150 cm above the ground. Furthermore, during the field-testing stage, our system proved to be efficient, and it is planned that in the future it will be used to determine attenuation curves from the measurement and recording of RSSI in other greenhouses of different crops in the 868/915 and 2400 MHz bands. A dataset of the radio wave attenuation can be registered in tomato greenhouse and can be found at the following link [52]. Likewise, for ease of deployment and use by actors in agricultural production, the authors are planning to design an integrated hardware and software solution based on our system.

Author Contributions: Conceptualization, M.D.-H., J.A.H.-T., F.G.-M., D.C.-P. and A.C.-P. data curation, D.C.-P. and J.A.H.-T. formal analysis, D.C.-P., M.D.-H., J.A.H.-T., F.G.-M. and A.C.-P. funding acquisition, D.C.-P. and A.C.-P. investigation, D.C.-P., M.D.-H., J.A.H.-T., F.G.-M. and A.C.-P. methodology, D.C.-P., M.D.-H., J.A.H.-T. and A.C.-P. project administration, D.C.-P., M.D.-H. and A.C.-P. resources, D.C.-P. software, D.C.-P. and A.C.-P. supervision, M.D.-H. and F.G.-M. validation, M.D.-H. and A.C.-P. writing—original draft, D.C.-P. and A.C.-P. writing—review and editing, D.C.-P. and A.C.-P. All authors have read and agreed to the published version of the manuscript.

Funding: This research received fund by the Ibero-American Postgraduate University Association (AUIP).

Data Availability Statement: http://data.mendeley.com/datasets/nhk3gs7gmm/1 (accessed on 8 September 2021).

Conflicts of Interest: The authors declare no conflict of interest.

References

1. Rangwani, D.; Sadhukhan, D.; Ray, S.; Khan, M.K.; Dasgupta, M. An improved privacy preserving remote user authentication scheme for agricultural wireless sensor network. *Trans. Emerg. Telecommun. Technol.* **2021**, *32*, e4218. [CrossRef]
2. Cama-Pinto, D.; Damas, M.; Holgado-Terriza, J.; Gómez-Mula, F.; Calderin-Curtidor, A.; Martínez-Lao, J.; Cama-Pinto, A. 5G Mobile Phone Network Introduction in Colombia. *Electronics* **2021**, *10*, 922. [CrossRef]
3. Mentsiev, A.U.; Gatina, F.F. Data analysis and digitalisation in the agricultural industry. *IOP Conf. Series Earth Environ. Sci.* **2021**, *677*, 32101. [CrossRef]
4. Azman, A.S.; Lee, M.Y.; Subramaniam, S.K.; Feroz, F.S. Novel Wireless Sensor Network Routing Protocol Performance Evaluation using Diverse Packet Size for Agriculture Application. *Int. J. Integr. Eng.* **2021**, *13*, 16–28. [CrossRef]
5. Vanishree, K.; Nagaraja, G.S. Emerging Line of Research Approach in Precision Agriculture: An Insight Study. *Int. J. Adv. Comput. Sci. Appl.* **2021**, *12*. [CrossRef]
6. Peng, Y.; Xiao, Y.; Fu, Z.; Dong, Y.; Zheng, Y.; Yan, H.; Li, X. Precision irrigation perspectives on the sustainable water-saving of field crop production in China: Water demand prediction and irrigation scheme optimization. *J. Clean. Prod.* **2019**, *230*, 365–377. [CrossRef]
7. Caicedo-Ortiz, J.G.; De-La-Hoz-Franco, E.; Ortega, R.M.; Piñeres-Espitia, G.; Combita-Niño, H.; Estévez, F.; Cama-Pinto, A. Monitoring system for agronomic variables based in WSN technology on cassava crops. *Comput. Electron. Agric.* **2018**, *145*, 275–281. [CrossRef]
8. Caicedo Ortiz, J.G.; Acosta Coll, M.A.; Cama-Pinto, A. WSN deployment model for measuring climate variables that cause strong precipitation. *Prospectiva* **2015**, *13*, 106–115. [CrossRef]
9. Miao, Y.; Zhao, C.; Wu, H. Non-uniform clustering routing protocol of wheat farmland based on effective energy consumption. *Int. J. Agric. Biol. Eng.* **2021**, *14*, 142–150. [CrossRef]
10. Razafimandimby, C.; Loscri, V.; Vegni, A.M.; Neri, A. Efficient Bayesian Communication Approach for Smart Agriculture Applications. In Proceedings of the 2017 IEEE 86th Vehicular Technology Conference (VTC-Fall), Toronto, ON, Canada, 24–27 September 2017; Institute of Electrical and Electronics Engineers (IEEE): New York, NY, USA, 2017; pp. 1–5.
11. Salim, C.; Mitton, N. K-predictions based data reduction approach in WSN for smart agriculture. *Computing* **2020**, *103*, 509–532. [CrossRef]
12. Wu, H.; Miao, Y.; Li, F.; Zhu, L. Empirical Modeling and Evaluation of Multi-Path Radio Channels on Wheat Farmland Based on Communication Quality. *Trans. ASABE* **2016**, *59*, 759–767. [CrossRef]
13. Cama-Pinto, A.; Gil Montoya, F.; Gómez-López, J.; García-Cruz, A.; Manzano-Agugliaro, F. Wireless surveillance sytem for greenhouse crops. *DYNA* **2014**, *81*, 164. [CrossRef]
14. Montoya, F.G.; Gomez, J.; Manzano-Agugliaro, F.; Cama, A.; García-Cruz, A.; De La Cruz, J.L. 6LoWSoft: A software suite for the design of outdoor environmental measurements. *J. Food Agric. Environ.* **2013**, *11*, 2584–2586.
15. Hsiao, S.-J.; Sung, W.-T. A Study on Using a Wireless Sensor Network to Design a Plant Monitoring System. *Intell. Autom. Soft Comput.* **2021**, *27*, 359–377. [CrossRef]
16. Xuanrong, P.; Tingdong, Y.; Yuesheng, W. Research and design of precision irrigation system based on artificial neural network. In Proceedings of the 2018 Chinese Control and Decision Conference (CCDC), Shenyang, China, 9–11 June 2018; Institute of Electrical and Electronics Engineers (IEEE): New York, NY, USA, 2018; pp. 3865–3870.
17. Zapata-Sierra, A.J.; Cama-Pinto, A.; Montoya, F.G.; Alcayde, A.; Manzano-Agugliaro, F. Wind missing data arrangement using wavelet based techniques for getting maximum likelihood. *Energy Convers. Manag.* **2019**, *185*, 552–561. [CrossRef]

18. Zhang, H.; Li, H. Node Localization Technology of Wireless Sensor Network Based on RSSI Algorithm. *Int. J. Online Eng.* **2016**, *12*, 51–57. [CrossRef]
19. Azmi, N.; Kamarudin, L.; Zakaria, A.; Ndzi, D.; Rahiman, M.; Zakaria, S.; Mohamed, L. RF-Based Moisture Content Determination in Rice Using Machine Learning Techniques. *Sensors* **2021**, *21*, 1875. [CrossRef]
20. Piñeres-Espitia, G.; Cama-Pinto, A.; De La Rosa Morrón, D.; Estevez, F.; Cama-Pinto, D. Design of a low cost weather station for detecting environmental changes. *Espacios* **2017**, *38*, 13.
21. Foerster, A.; Udugama, A.; Görg, C.; Kuladinithi, K.; Timm-Giel, A.; Cama-Pinto, A. A Novel Data Dissemination Model for Organic Data Flows. In Proceedings of the International Conference on Mobile Network and Management, Santander, Spain, 16–18 September 2015; Springer: Berlin/Heidelberg, Germany, 2015; pp. 239–252.
22. Cama-Pinto, A.; Gil Montoya, F.; Gómez, J.; De La Cruz, J.L.; Manzano-Agugliaro, F. Integration of communication technologies in sensor networks to monitor the Amazon environment. *J. Clean. Prod.* **2013**, *59*, 32–42. [CrossRef]
23. Farooqui, N.A.; Tyagi, A. Data Mining and Fusion Techniques for Wireless Intelligent Sensor Networks. In *Handbook of Wireless Sensor Networks: Issues and Challenges in Current Scenario's*; Springer: Berlin/Heidelberg, Germany, 2020; pp. 592–615.
24. Montoya, F.G.; Gómez, J.; Cama, A.; Zapata-Sierra, A.; Martínez, F.; De La Cruz, J.L.; Manzano-Agugliaro, F. A monitoring system for intensive agriculture based on mesh networks and the android system. *Comput. Electron. Agric.* **2013**, *99*, 14–20. [CrossRef]
25. Maiolo, L.; Polese, D. Advances in sensing technologies for smart monitoring in precise agriculture. In Proceedings of the SENSORNETS 2021—Proceedings of the 10th International Conference on Sensor Networks, Vienna, Austria, 9–10 February 2021; pp. 151–158.
26. Sathish, C.; Srinivasan, K. An artificial bee colony algorithm for efficient optimized data aggregation to agricultural IoT devices application. *J. Appl. Sci. Eng.* **2021**, *24*, 927–936. [CrossRef]
27. Saiz-Rubio, V.; Rovira-Más, F. From Smart Farming towards Agriculture 5.0: A Review on Crop Data Management. *Agronomy* **2020**, *10*, 207. [CrossRef]
28. Subashini, M.M.; Das, S.; Heble, S.; Raj, U.; Karthik, R. Internet of Things based wireless plant sensor for smart farming. *Indones. J. Electr. Eng. Comput. Sci.* **2018**, *10*, 456–468. [CrossRef]
29. Abouzar, P.; Michelson, D.G.; Hamdi, M. RSSI-Based Distributed Self-Localization for Wireless Sensor Networks Used in Precision Agriculture. *IEEE Trans. Wirel. Commun.* **2016**, *15*, 6638–6650. [CrossRef]
30. Xu, L. Design of a RSSI Location System for Greenhouse Environment. *Int. J. Distrib. Sens. Netw.* **2015**, *11*, 525861. [CrossRef]
31. Cama-Pinto, D.; Damas, M.; Holgado-Terriza, J.A.; Gomez-Mula, F.; Cama-Pinto, A. Desarrollo de un sistema para medición y registro de RSSI en invernaderos. *Av. En Arquit. Y Tecnol. De Comput. Actas De Las Jorn. SARTECO* **2019**, 649–654. [CrossRef]
32. Li, T.; Zhang, M.; Ji, Y.H.; Sha, S.; Jiang, Y.Q.; Li, M.Z. Management of CO2 in a tomato greenhouse using WSN and BPNN techniques. *Int. J. Agric. Biol. Eng.* **2015**, *8*, 43–51. [CrossRef]
33. García, L.; Parra, L.; Jimenez, J.; Parra, M.; Lloret, J.; Mauri, P.; Lorenz, P. Deployment Strategies of Soil Monitoring WSN for Precision Agriculture Irrigation Scheduling in Rural Areas. *Sensors* **2021**, *21*, 1693. [CrossRef]
34. Aung, S.M.Y.; Pattanaik, K.K. Path Loss Measurement for Wireless Communication in Industrial Environments. In Proceedings of the 2020 International Conference on Computer Science, Engineering and Applications (ICCSEA), Gunupur, India, 13–14 March 2020; Institute of Electrical and Electronics Engineers (IEEE): New York, NY, USA, 2020; pp. 1–5.
35. Navarro, A.; Guevara, D.; Florez, G.A. An Adjusted Propagation Model for Wireless Sensor Networks in Corn Fields. In Proceedings of the 2020 XXXIIIrd General Assembly and Scientific Symposium of the International Union of Radio Science, Rome, Italy, 29 August–5 September 2020; Institute of Electrical and Electronics Engineers (IEEE): New York, NY, USA, 2020.
36. Pal, P.; Sharma, R.P.; Tripathi, S.; Kumar, C.; Ramesh, D. 2.4 GHz RF Received Signal Strength Based Node Separation in WSN Monitoring Infrastructure for Millet and Rice Vegetation. *IEEE Sens. J.* **2021**, *21*, 18298–18306. [CrossRef]
37. Wang, J.; Peng, Y.; Li, P. Propagation Characteristics of Radio Wave in Plastic Greenhouse. In Proceedings of the International Conference on Computer and Computing Technologies in Agriculture, Beijing, China, 27–30 September 2015; Springer: Berlin/Heidelberg, Germany, 2016; pp. 208–215.
38. Widodo, S.; Pratama, E.A.; Pramono, S.; Basuki, S.B. Outdoor propagation modeling for wireless sensor networks 2.4 GHz. In Proceedings of the 2017 IEEE International Conference on Communication, Networks and Satellite (Comnetsat), Semarang, Indonesia, 5–7 October 2017; Institute of Electrical and Electronics Engineers (IEEE): New York, NY, USA, 2017; pp. 158–162.
39. Cama-Pinto, A.; Espitia, G.D.P.; Caicedo, J.G.; Ramirez-Cerpa, E.; Betancur-Agudelo, L.; Gómez-Mula, F. Received strength signal intensity performance analysis in wireless sensor network using Arduino platform and XBee wireless modules. *Int. J. Distrib. Sens. Netw.* **2017**, *13*. [CrossRef]
40. Shue, S.; Johnson, L.E.; Conrad, J.M. Utilization of XBee ZigBee modules and MATLAB for RSSI localization applications. In Proceedings of the SoutheastCon 2017, Concord, NC, USA, 30 March–2 April 2017; Institute of Electrical and Electronics Engineers (IEEE): New York, NY, USA, 2017.
41. Cama-Pinto, D.; Damas, M.; Holgado-Terriza, J.A.; Gómez-Mula, F.; Cama-Pinto, A. Path Loss Determination Using Linear and Cubic Regression Inside a Classic Tomato Greenhouse. *Int. J. Environ. Res. Public Health* **2019**, *16*, 1744. [CrossRef] [PubMed]
42. Van Herbruggen, B.; Jooris, B.; Rossey, J.; Ridolfi, M.; Macoir, N.; Van Den Brande, Q.; Lemey, S.; De Poorter, E. Wi-pos: A low-cost, open source ultra-wideband (UWB) hardware platform with long range sub-GHZ backbone. *Sensors* **2019**, *19*, 1548. [CrossRef]

43. Bezunartea, M.; Wang, C.; Braeken, A.; Steenhaut, K. Multi-radio Solution for Improving Reliability in RPL. In Proceedings of the 2018 IEEE 29th Annual International Symposium on Personal, Indoor and Mobile Radio Communications (PIMRC), Bologna, Italy, 9–12 September 2018; Institute of Electrical and Electronics Engineers (IEEE): New York, NY, USA, 2018; pp. 129–134.
44. Texas Instruments—Descripción CC2538. Available online: http://www.ti.com/product/CC2538/description (accessed on 21 July 2021).
45. Gomez, J.; Villar, E.; Molero, G.; Cama, A. Evaluation of high performance clusters in private cloud computing environments. In *Distributed Computing and Artificial Intelligence*; Springer: Berlin/Heidelberg, Germany, 2012; pp. 305–312.
46. ERCIM News. Contiki: Bringing IP to Sensor Networks. Available online: https://ercim-news.ercim.eu/en76/rd/contiki-bringing-ip-to-sensor-networks (accessed on 21 July 2021).
47. Cama-Pinto, D.; Damas, M.; Holgado-Terriza, J.A.; Arrabal-Campos, F.M.; Gómez-Mula, F.; Martínez-Lao, J.A.M.; Cama-Pinto, A. Empirical Model of Radio Wave Propagation in the Presence of Vegetation inside Greenhouses Using Regularized Regressions. *Sensors* **2020**, *20*, 6621. [CrossRef] [PubMed]
48. Staudemeyer, R.C.; Pöhls, H.C.; Wójcik, M. What it takes to boost Internet of Things privacy beyond encryption with unobservable communication: A survey and lessons learned from the first implementation of DC-net. *J. Reliab. Intell. Environ.* **2019**, *5*, 41–64. [CrossRef]
49. Dunkels, A.; Gronvall, B.; Voigt, T. Contiki—A lightweight and flexible operating system for tiny networked sensors. In Proceedings of the 29th Annual IEEE International Conference on Local Computer Networks, Tampa, FL, USA, 16–18 November 2004; Institute of Electrical and Electronics Engineers (IEEE): New York, NY, USA, 2004.
50. Dunkels, A.; Österlind, F.; He, Z. An adaptive communication architecture for wireless sensor networks. In Proceedings of the SenSys'07—Proceedings of the 5th ACM Conference on Embedded Networked Sensor Systems, Sydney, Australia, 6–9 November 2007; Machinery: New York, NY, USA, 2007; pp. 335–349.
51. Vougioukas, S.; Anastassiu, H.; Regen, C.; Zude, M. Influence of foliage on radio path losses (PLs) for wireless sensor network (WSN) planning in orchards. *Biosyst. Eng.* **2013**, *114*, 454–465. [CrossRef]
52. Raheemah, A.; Sabri, N.; Salim, M.S.; Ehkan, P.; Ahmad, R.B. New empirical path loss model for wireless sensor networks in mango greenhouses. *Comput. Electron. Agric.* **2016**, *127*, 553–560. [CrossRef]
53. Cama-Pinto, D.; Holgado-Terriza, J.A.; Damas, M.; Gómez-Mula, F.; Cama-Pinto, A. Tomato Greenhouse Measurement of RSSI in Almeria Spain. Available online: https://data.mendeley.com/datasets/nhk3gs7gmm/1 (accessed on 8 September 2021).
54. Zennaro, M.; Bagula, A.; Gascon, D.; Noveleta, A.B. Long distance wireless sensor networks: Simulation vs. reality. In Proceedings of the 4th ACM Workshop on Networked Systems for Developing Regions, NSDR'10, San Francisco, CA, USA, 15 June 2010.

 *inventions*

Article

Expanding the Level of Technological Readiness for a Low-Cost Vertical Hydroponic System

Juan D. Borrero

Department of Business and Marketing, Huelva University, 21071 Huelva, Spain; jdiego@uhu.es

Abstract: Climate and social changes are deeply affecting current agro-food systems. Unsustainable agricultural practices and the low profitability of small farmers are challenging the agricultural development of rural areas. This study aims to develop a novel, modular and low-cost vertical hydroponic farm system through reviews of the patented literature, research literature and variants of commercial products. After a detailed conceptualization process, a prototype was fabricated and tested at my university to validate its technology readiness level (TRL). The outcomes supported the usability and performance of the present utility model but highlighted several changes that are necessary before it can pass to the next TRL. This study shows that the prototype has the potential to not only solve food sovereignty but also to benefit society by advancing the innovations in food production and improving quality of life.

Keywords: vertical farming; indoor farming; low-cost vertical farming; small farming; vertical hydroponics; technology readiness level; utility model

Citation: Borrero, J.D. Expanding the Level of Technological Readiness for a Low-Cost Vertical Hydroponic System. *Inventions* **2021**, *6*, 68. https://doi.org/10.3390/inventions6040068

Academic Editors: Francisco Manzano Agugliaro, Esther Salmerón-Manzano and Konstantinos G. Arvanitis

Received: 25 July 2021
Accepted: 11 October 2021
Published: 14 October 2021

Publisher's Note: MDPI stays neutral with regard to jurisdictional claims in published maps and institutional affiliations.

1. Introduction

By 2025, the world's population will number 9.7 billion [1]. Consequently, the agricultural sector will need to produce more food [2]. However, crop production is currently affected by many factors such as climate change, a lack of water and arable land shortages [3,4]. Despite these issues, agricultural productivity needs to be improved.

Agriculture is responsible for 60% of the total water use in Europe [5] and 70% in the world [6]. In addition, agriculture for food uses large amounts of fertilizer and pesticide: the European Union (EU) uses more than 11 million tons of fertilizer each year [7] and the total EU pesticide sales volume was around 350,000 tons per year on average during the 2011–2019 period [8]. We need to protect our environment and natural resources, but excessive fertilizer use can pollute the surface and groundwater, while the use of pesticides damages both human health and the environment.

The need for a sustainable and cheaper source of food production has led to increasing research on several vertical farm systems [9–13]. To meet the increasing demand for food in concentrated areas or big cities, the research on last-mile consumption has also recently increased [14,15]. If fresh food is produced locally, then it is not necessary to travel far to obtain it, which reduces carbon footprint [16].

Currently, there are numerous farming techniques used to reduce water consumption, environmental impacts, and space for crops. Vertical farming innovations provide both food security and ensure environmental sustainability in small sites [17,18]. Vertical farms could also provide young farmers, constrained by land or credit access [19], with the opportunity to run a centrally located farm while pursuing a university degree [20].

The technology developed in this paper has three general aims: the efficient use of natural resources and space, sustainability, and efficiencies in the food value chain which reduce the carbon footprint by producing food locally.

Vertical hydroponic farming is a combination of vertical production and hydroponic methods that are proven to be useful [9–13].

Hydroponics is a method of growing crops without soil by using mineral nutrients in a water solution [21]. Various hydroponic techniques, such as a drip system, nutrient film, deep flow, or aeroponics, are available [21]. Different crops can be grown in a vertical hydroponic system, making indoor farming possible [22].

Several types of commercial vertical hydroponic system products such as A-frame, Zig-zag tower, or ZipGrow are available, as well as vertical hydroponic plant factories (e.g., Aerofarms, Plenty, or Bowery) which produce vegetables in urban areas [23–25]. Although these systems are helping to meet the local demand for food, they remain costly and complex for small growers.

The aim herein is to build and test a modular and low-cost invention that may be utilized by citizens or small farmers, indoors or outdoors, in urban or rural areas. In this way, every home could be a farm, and a small farm based around a city could supply enough food to the local villagers. These are the major motivations for this paper.

Although the literature on vertical farming focuses on various aspects, no studies promote a local food sovereignty solution through a scalable low-cost structure. While some research emphasizes technology, none provides a detailed solution for modular structures [26–33].

Moreover, these studies do not include detailed designs that could be useful for other researchers to use to replicate or improve this innovation. To my knowledge, this is the first implemented vertical hydroponic system with a Technology Readiness Level (TRL) analysis.

In this paper, a specific methodology is implemented to develop a utility model, i.e., a system similar to the patents which provide protection via minor improvements [34], before passing to the engineering stage.

Thus, the present invention consists of a system based on a modular support structure, which is intended for the hydroponic cultivation of a wide variety of crops. Specifically, the invention focuses on the creation of a multi-level structure with a fully adjustable height. This structure is modular and can be used in both large and small areas, i.e., in greenhouses, both indoors or outdoors, in gardens, or on balconies.

The remainder of the paper is presented as follows: Section 2 covers a review of vertical farms in patents, describes the TRL methodology, and provides a comprehensive description of the vertical farming system designed and developed. Section 3 presents the results of the implemented system and compares them with some commercial and research solutions. Finally, the conclusions are presented in Section 4.

2. Materials and Methods

In the following section, the most current patents related to hydroponic vertical systems, which are already published, are reviewed and critically evaluated. Secondly, the design of the low-cost invention is described, which is considered a utility model, and, finally, the methodology is presented.

2.1. Integrative Research Review

In patent document EP 2 904 894 A1 [35], a modular hydroponic growing system with hanging units in the form of geotextile bags is shown. This same structure serves as a support for irrigation and drainage systems. The solution is designed mainly for the planting of small vegetables both in greenhouses and indoors.

In patent document US2011/0067301 A1 [36], a hydroponic growing system is shown to provide a continuous flow of nutrients to plants. The system is made up of large-diameter tubes comprising a plurality of cutouts and interconnected to each other with a slight slope. A pump supplies the nutrients through the upper part of the structure, collecting the drainage water from the lower part. As in the previous document, the system is designed for small crops. Furthermore, this solution cannot be made modular.

Patent document US2012/0066972 A1 [37] shows a vertical planting system with the plants in bags hanging from the structure. In the lower part of it, there is a water storage tank with a pump capable of supplying water to the upper part of the structure. In this case, a modular system is not possible either.

Patent document US 7,055,282 B2 [38] shows a vertical hydroponic cultivation device for greenhouses. The system consists of a tower comprised of modules with hollow interiors, along with inclined support cups where crops are planted. As in the previous cases, this system is designed for small crops.

None of the patents mentioned are versatile enough to be used with both small crops such as greens (e.g., lettuce, kale, or basil), medium crops such as strawberries, or larger crops such as blueberries. They are all only suitable for crops of small sizes and weights, while the present invention is capable of supporting the growth of all of these crops.

It is also important to note that, although all the aforementioned patents describe the increase in plant density (number of individuals per unit ground area), none of them suggest or disclose that the modules can be assembled together. The present invention allows for the coupling of modules with telescopic legs to guarantee water drainage throughout the entire system. In addition, no solution was designed for cultivation in rural farms, where simple structures such as those presented in this invention were needed, structures which made it possible to join many modules.

Another advantage of the present invention is that it allows the use of commercial substrates (e.g., perlite, fiber) and containers (e.g., sacks, bags, and pots). Therefore, the structure is compatible with the commercial components from hydroponics, fertigation (e.g., drip irrigation), or drainage (e.g., gutters), while the patents mentioned need to adapt the commercial components to the patented structure.

Finally, the present invention can be used for both hydroponics and aeroponics, for which the invention comprises interchangeable components.

Taking into account the aforementioned points, it can be said that there are a multitude of solutions. However, a system with a high degree of versatility that can be used in different environments and situations is essential and, so far, this system has not been demonstrated by the systems and structures put forward to date. In addition, because hydroponics are used more and more in all types of plantations, a system adaptable to any type of crop is needed. To solve this technical problem and the needs raised above, the invention that is described below makes it possible to obtain a vertical system for totally versatile hydroponic cultivation, which can be used in a great variety of situations and crops.

2.2. System Design

Although it was not the objective of this study and the data are not presented since the experiment was not concluded, a pilot test was conducted that covered other interventions such as studying the behavior of small crops. Strawberries (*Fragaria* spp.) were planted in the first module, and lettuce (*Lactuca sativa*) and purple cabbage (*Brassica oleracea* var. *capitata* f. *rubra*) were planted in the second. The heights of the different levels were studied with the aim of optimizing the farming labor required for these crops. The degree of the influence of shaded areas in the different levels of the structure and their influence on the final production yield were evaluated, and two AC-powered LED strip lights were located in the module that received the most shade to assess the influence on the crop development.

The construction of two vertical hydroponic modules with fixed heights were commissioned for the pilot test. According to the ES 1 242 949 U utility model [39], the system (Figure 1a,b) was made up of two floor stands (1) on which the rest of the structure was attached. The main mission of the floor stand was to prevent the crop from tipping over, as well as to distribute the weight of the structure on the ground.

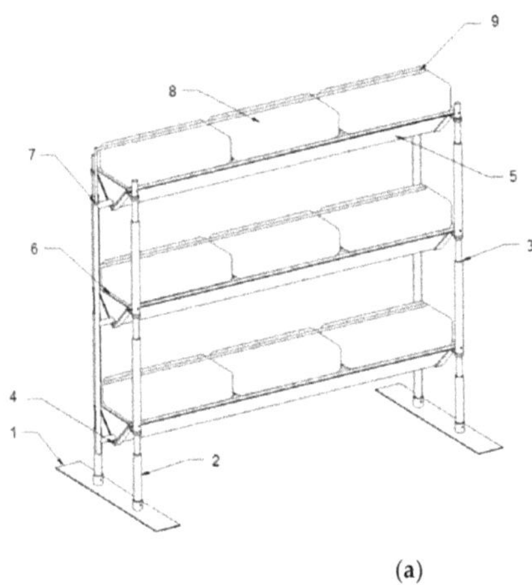

(a)

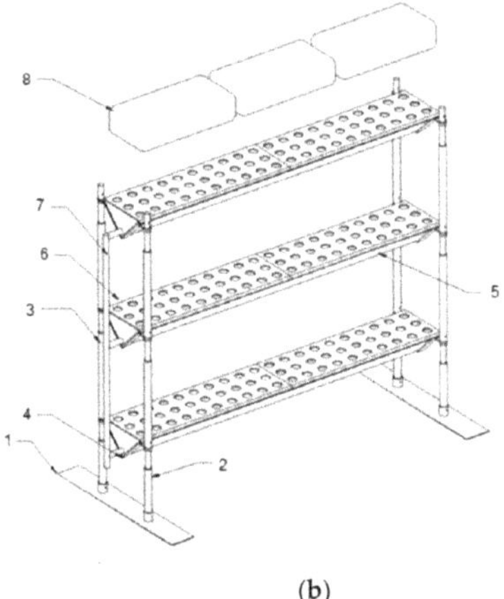

(b)

Figure 1. View of the system with all the components that make up the structure. With substrate sacks (**a**) Without substrate sacks (**b**).

On the floor stand, two telescopic legs (Figure 2b (2)) of a short length (30–40 cm) are installed, made up of two parts: the lower part (A) consists of a hollow tube with an internal thread, while the upper part (B) is a tube with an external thread. By spinning both parts together, the desired height can be achieved at each point.

To be able to attach the telescopic legs to each other or to the floor stand, there are two holes at each of their ends through which a screw (see Figure 2a between (2) and (3)) is installed, thus joining the two pieces. In addition, in the upper part of the telescopic leg. there is a stop that supports an elevated gutter.

The telescopic legs of the upper levels (Figures 1 and 2a (3)) are configured in a similar way to those of the lower level, with the only difference being that the tube with the lower thread has a greater length, so that a greater separation between cultivation levels is achieved.

The following elements are the supports (Figures 1 and 2a (4)) which are responsible for joining the telescopic legs to the drainage channel. These supports are V-shaped with two rings at the ends which are inserted into the telescopic legs.

One of the main elements which makes up the system is the drainage gutter (Figure 2d (5)). It is a V-shaped Polyvinyl chloride (PVC) channel in charge of collecting the excess water from the plant and directing it to the end of the system where it is guided into a tank. The gutters can be coupled longitudinally by means of reinforcements (13) or clamps (see Figure 2e). The tubes (Figure 1 (7) and Figure 3 (12)) are pipes which lead the drainage water from the lower end of the drainage gutter to the lower part of the structure where it is stored in a container or tank.

There is a metal lattice (Figure 2d (6)) that supports the substrate sacks (8) attached to the drainage gutter.

Additionally, the system has a drip irrigation tape (Figure 1a (9)) that is arranged on top of the substrate sacks.

The structure described above can also be adapted to aeroponic crops using plastic tubes (Figure 3) with a plurality of holes in their upper part where the crops are housed. These tubes serve as drainage gutters (11). In this case, straight sheets (10) must be used as supports with two holes at each of their ends (Figure 2f (10), instead of Figure 2a (4)). Although not designed for this utility model, the seeds are "planted" in pieces of foam stuffed into tiny pots, which are exposed to the light at one end, leaving the roots to dangle in the air, where they are periodically sprayed by aerosol-generating nozzles inside the pipes.

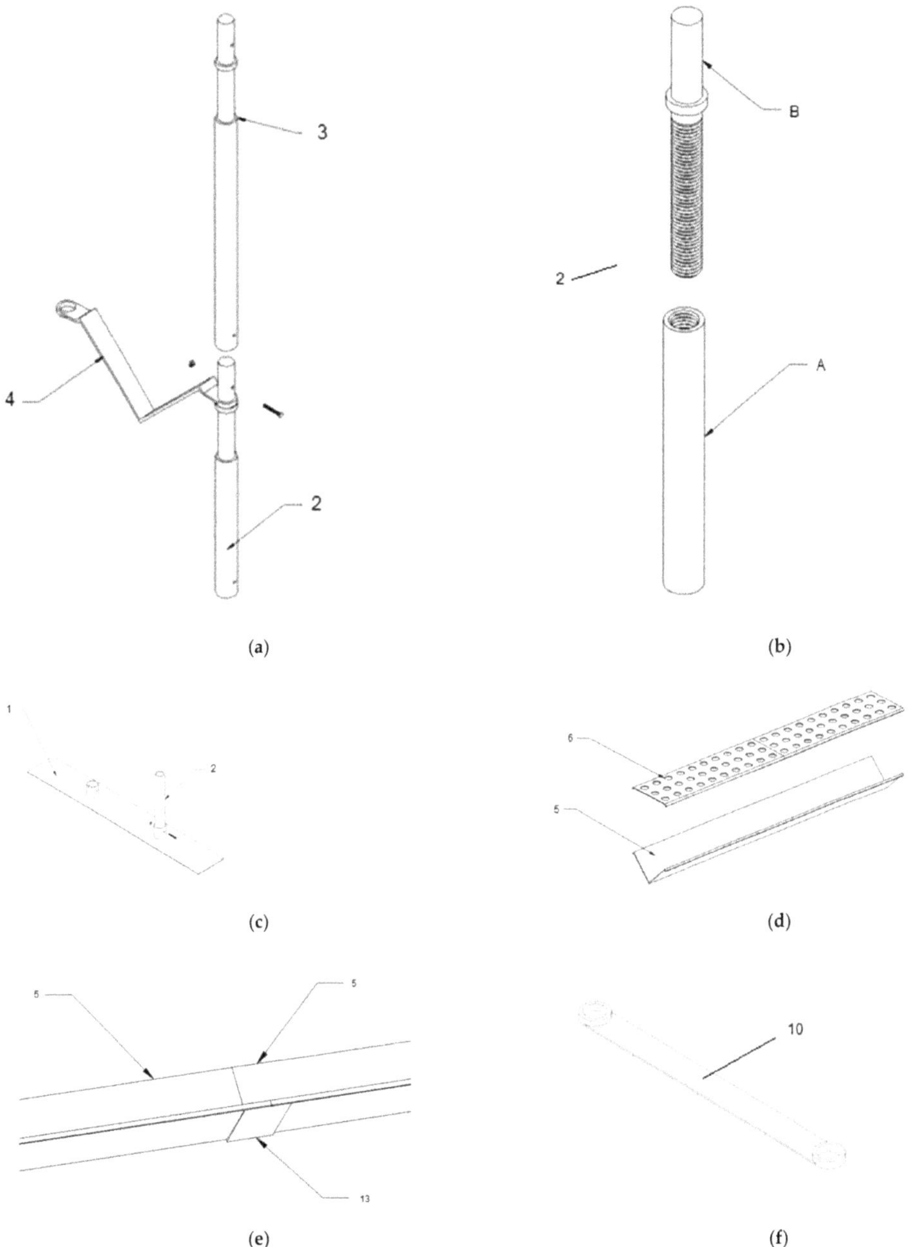

Figure 2. Components in detail. From left to right and from top to bottom: (**a**) Junction of two levels. (**b**) Telescopic leg. (**c**) Splicing of the legs with the floor stand. (**d**) Configuration of a drainage gutter and a metal lattice. (**e**) The splice between two gutters. (**f**) Component for aeroponic cultivation.

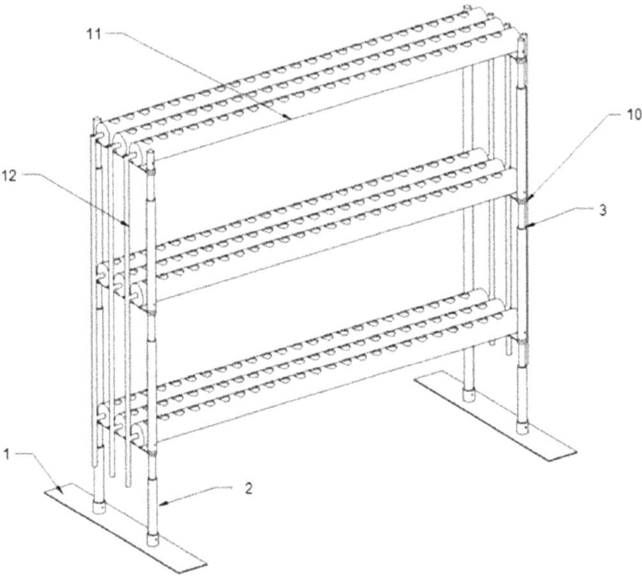

Figure 3. Detail of the adaptation to an aeroponic system.

Using only the lower level, the system can accommodate large crops which are planted in pots, such as blueberries (Figure 4).

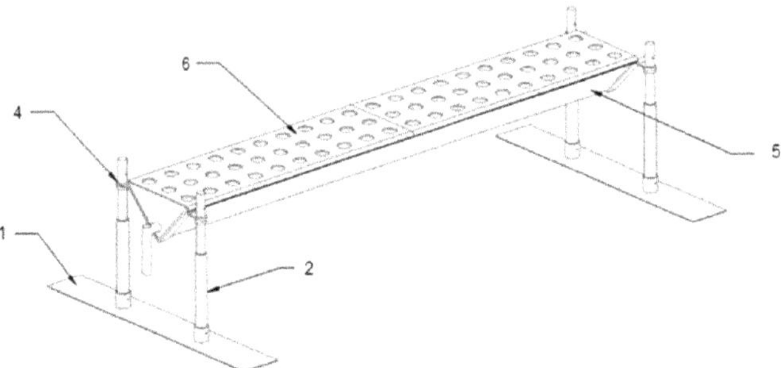

Figure 4. Representation in a free perspective of the adaptation of the system to large crops.

Another feature of this invention is that it is designed to be able to couple the modules that are necessary, thus it is able to cover long distances (Figure 5). In this case, the height of each level must be correctly adjusted, using the telescopic legs and ensuring the hermetic seal of the gutters (see Figure 2e (13)) to achieve a slight slope that runs through the entire row of a crop.

Furthermore, the use of LED lights is possible as they can stimulate the growth of crops in locations with a few hours of daylight. LEDs are solid-state elements (they do not require ballasts) which emit very little heat and can be located very near to the crops [40–42].

Aside from these advantages, the low energy consumption, high durability (the average useful life is 50,000 h), instantaneous ignition, different types of LEDs offered, the fact that they can be contained in a single strip, the low cost, and the improved energy efficiency of LEDs are factors which justify their adoption in agriculture.

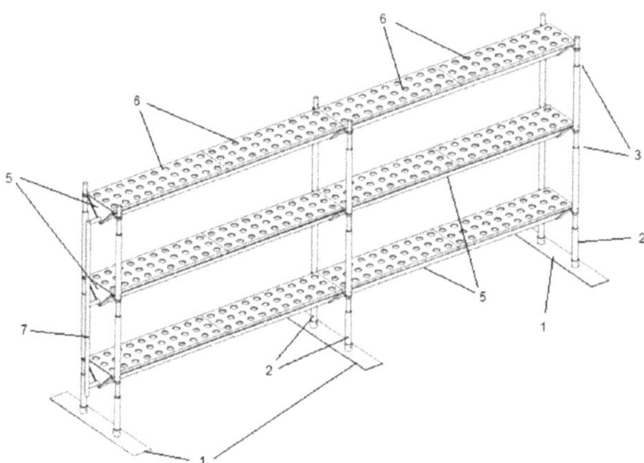

Figure 5. Free perspective representation of the adaptation to dimensions and heights by joining several modules.

The wavelengths of light corresponding to photochemical processes are in the range from 400 to 520 nanometers (nm) corresponding to the visible spectrum, and comprise violet, blue, and green light, which have a strong influence on vegetative growth and photosynthesis; and 610 and 720 nm, also corresponding to the visible spectrum, comprising red light, which stimulates the vegetative growth, photosynthesis, flowering, and germination by means of the crops' photosensitive pigments [43].

Naznin et al. [44] investigated the growth of strawberry plants with different red:blue LED ratios (5:1, 10:1; 19:1) versus High Pressure Sodium vapor (HPS) lights, obtaining better results in all parameters (leaf number, runner number, inflorescences number, crown number, length of flowering, stems per plant, and dry mass) with LED lights.

Other research carried out on strawberries by Hanenberg, Janse and Verkerke [45] also found better results for brix and vitamin C production and in sensory tests with LED lighting near the leaves or fruits. According to these authors, the illumination of the leaves resulted in the highest brix levels, while the illumination of the fruits resulted in the highest levels of vitamin C. These observations were in agreement with Gautier et al. [46].

According to Mochizuki et al. [47], the short-distance lighting with LEDs in strawberry production, irradiated on the underside of the leaves with blue LEDs, improved assimilation in young leaves compared to the use of red LEDs.

As the province of Huelva (Spain) has a good number of hours of daylight [48], the strawberry plant already obtained the required luminosity; therefore, red LEDs were, in this case, more necessary than blue ones to promote flowering and achieve a better performance.

Thus, two LED strips were used, one with red and blue 3:1 (3Red:1Blue) and the other with 5:1 (5Red:1:Blue) color combinations. These strips were placed on the module that received the most shade, placing the 5:1 strip on the upper level (red marking) and the 3:1 (blue marking) on the intermediate level (Figure 6).

Compared to the technologies and commercial solutions known in the agricultural sector, the structure on which the invention is based allows a modular system for hydroponic cultivation to be achieved which is totally versatile and adaptable to a wide variety of crops and situations, cultivating crops in several vertical segments, and thus increasing the plant density.

2.3. Technological Readiness Level (TRL)

TRLs are the constitutive scales of a method to estimate the technical maturity of different types of technologies. The main objective of using the TRL is to help make decisions related to technology development.

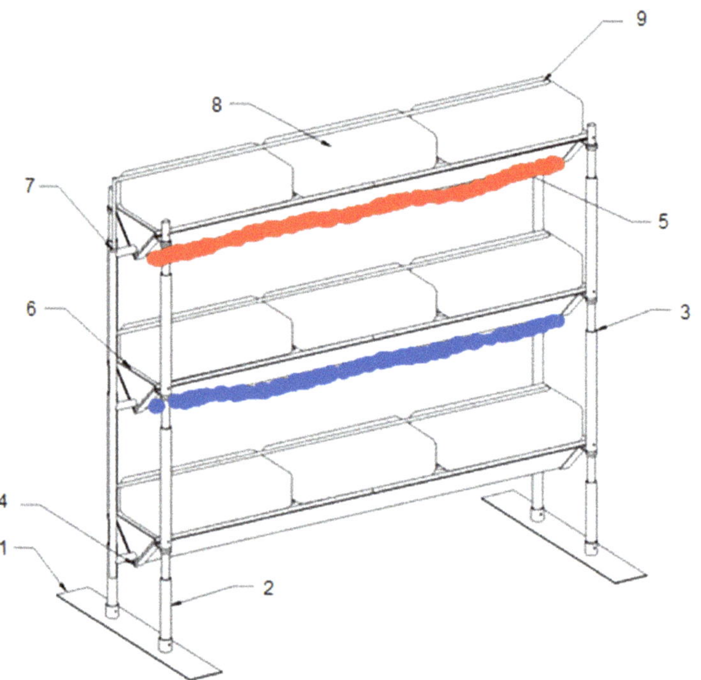

Figure 6. Installed LEDs in the vertical farm system.

This concept was developed at the National Aeronautics and Space Administration (NASA) during the 1970s for the space programs [49] and was subsequently formally adopted worldwide [50]. In the 2000s, the US Department of Defense used the scale for acquisitions [51]. In 2008, the scale was also used by the European Space Agency (ESA) [52]. In 2013, the TRL scale was formalized through the ISO 16290 [53].

In Europe, TRLs are determined using nine levels [54]:

- TRL 1. This is the lowest level of technological maturity at which scientific research begins to translate into applied research.
- TRL 2. Once fundamentals are verified, practical applications are devised. The examples are limited to speculative studies or utility models.
- TRL 3. In this stage, the product development starts. The work proceeds to the experimental phase to verify that the concept operates as expected.
- TRL 4. This level is considered as "low-fidelity" and determines if the individual elements could work as a system (Systems Readiness Level, SRL) [55].
- TRL 5. The basic technology components are integrated in a "high-fidelity" lab-scale system.
- TRL 6. The engineering development begins. The prototype must be able to perform all the functions required for a real environment. This represents an important step in demonstrating the maturity of a technology.
- TRL 7. This step requires the demonstration of the prototype in a real situation. The final design is practically complete.
- TRL 8. This TRL constitutes the end of system development, with testing in its final form and under the expected conditions.
- TRL 9. The technology is ready and the commercial manufacturing process can begin.

The approach of this paper is based on the transfer of a utility model from TRL 2 to TRL 4.

3. Results and Discussion

In the section that follows, the performance and functionality of the utility model ES 1 242 949 U [39] are validated, the plant density is determined, and the costs of the system are compared with those of conventional hydroponics.

The pilot was developed inside one of the greenhouses of the experimental farm at the University of Huelva (Andalusia, Spain) during the months from October 2020 to June 2021 (Figure 7).

Figure 7. Experimental site for the pilot test.

3.1. Dimensions of the Modular System Built

The modular system was made up of a galvanized iron structure with two parallel floor stands at each end, on which three levels are supported. Each level is made up of a triangular structure, which serves to support the substrate sacks, two per level, and collect the drainage water. Each module has dimensions of 220 × 170 × 30 cm (Figure 8).

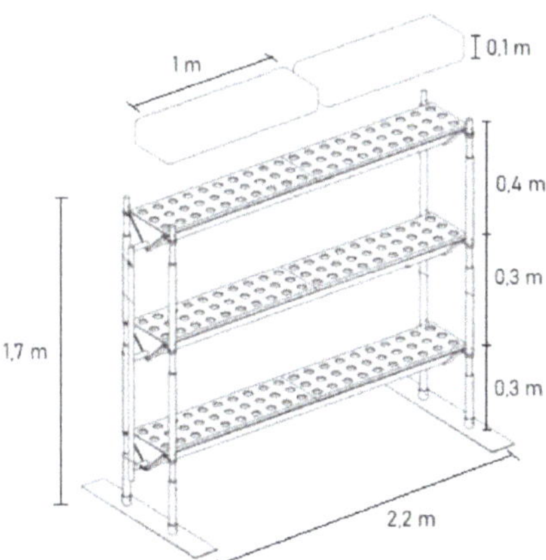

Figure 8. Dimensions of the vertical farm module system.

By carefully studying the heights at which the different levels were located, it was detected that the lower level, although higher than in a soil crop, was too low for the operator to easily pick fruit. In the same way, between the lower level and the intermediate one, the available height was 30 cm which, when subtracting the height of the sack, forced the plants (small crops) to grow up to 20 cm high. The height between the intermediate and upper level (40 cm) seemed to be optimal for the crop and its subsequent harvest. In addition, this intermediate level was at a height of 60 cm from the ground, which was ideal for carrying out management tasks. Finally, the upper level, although at an adequate height, made harvesting difficult for those operators of short stature.

To solve this technical problem, it was recommended to incorporate the telescopic legs indicated in the design of the utility model, and to replace the galvanized iron with aluminum and PCV.

3.2. LED Technology and Shading Influence

Using LED lights as supplementary lighting, with the addition of a single daily work cycle (six hours of light phase and two of dark phase) until the beginning of spring, I was able to visually verify that artificial lighting increased both the growth rate of the strawberry plants and their yield. Although more trials and repetitions are needed, provisional results show that the treatment with LED lights reduced the vegetative cycle of the plant by 5 days, from 60 to 55 days.

Due to how the modules were arranged inside the greenhouse (west–east), a slight shading was observed in some hours of the day which was eliminated with the use of artificial lighting.

3.3. Production and Costs

In a conventional macrotunnel (a simple structure that allows coverage of a large area where several rows of plants are grown in a controlled and protected environment), 12 plants were cultivated in each substrate sack, with four rows of plants separated by 100 cm (Figure 9a) between each. Instead, in this pilot (Figure 9b), the number of plants per sack was 13 and the corridors were narrowed by 10 cm to achieve a greater space for the structure and assure the optimal separation of the rows for harvesting so that they operated successfully (70–160 cm).

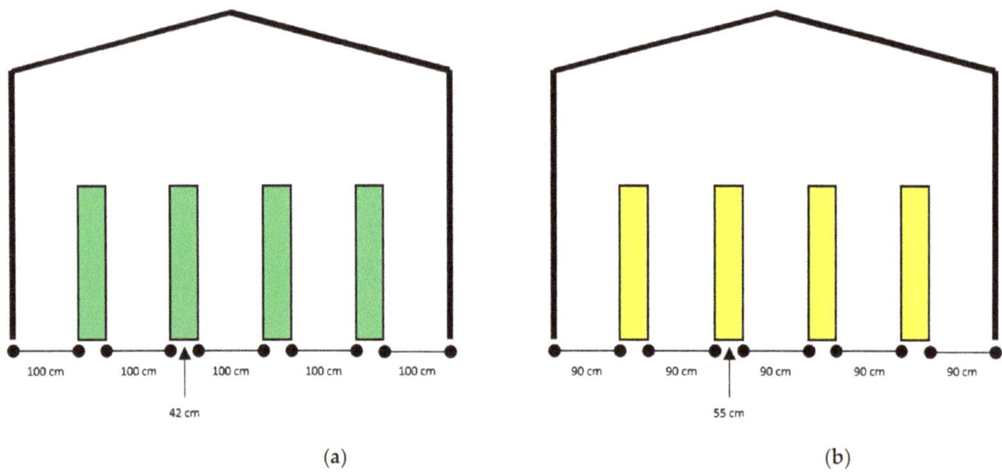

Figure 9. Arrangement of crop rows in the macrotunnel. 100 cm wide pathways (**a**) and 90 cm wide pathways (**b**).

Twenty-five macrotunnels of 7 m × 60 m can be used per hectare, so there can be 108 modules per tunnel. Although this does not imply more crop rows per macrotunnel, with these data, and assuming the same yield per plant (provisional results show that the yield production increased by 10%, from 900 g/plant to 1000 g/plant), the production per hectare can be increased by two to three times compared to conventional hydroponic systems (Table 1).

Table 1. Comparative plant density.

tunnel_width (m)	7
tunnel_length (m)	60
number_tunnels_per_ha	25
module_length (m)	2
number_modules_per_row_and_tunnel	27
number_rows_per_tunnel	4
number_modules_per_tunnel	108
number_substrate_bags_per_module_3_levels	6
number_substrate_bags_per_module_2_levels	4
number_substrate_bags_per_conventional_hydroponic_system	2
number_substrate_bags_per_tunnel_module_3_levels	648
number_substrate_bags_per_tunnel_module_2_levels	432
number_substrate_bags_per_tunnel_conventional_hydroponic_system	216
number_plants_per_tunnel_module_3_levels	8424
number_plants_per_tunnel_module_2_levels	5616
number_plants_per_tunnel_conventional_hydroponic_system	2592
number_plants_per_ha_module_3_levels	210,600
number_plants_per_ha_module_2_levels	140,400
number_plants_per_ha_conventional_hydroponic_system	64,800

The cost results (Table 2) show that when the pilot experiment was extended to cover one hectare, the total implementation costs were a little over twice as expensive as the conventional solution, but the cost per plant was much lower. If these data are compared with cost of the ZipGrow technique, the utility model proposed in this study is much cheaper [9]. According to ZipGrow [56], a fully equipped system of 92 m^2 for local production costs approximately USD 124,000 (fixed cost).

Table 2. Comparative system costs for a strawberry farm (EUR /ha).

	Conventional Hydroponic Solution	Three-Level Modules
Variable costs	22,680	69,660
Substrate sacks	16,200	48,600
Strawberry plants	6480	21,060
Fixed cost		
Hydroponic system	160,200	337,500
Total implementation cost	182,880	407,160
Total implementation cost per plant	28.22	19.33

Finally, Figures 10 and 11 show the images and provide further details of the pilot.

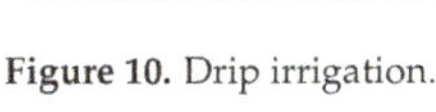

Figure 10. Drip irrigation.

Figure 11. LED lighting.

4. Conclusions

Having developed and tested the utility model, it can be affirmed that this innovation increases plant density and achieves a lower unit cost per plant.

The advantages provided by the pilot are: the linear square meters of the cultivated space and the number of plants grown per linear square meter are tripled, while the system is easy to install, easy to handle, makes for easy fruit picking, and enables the user great versatility to modify the structure according to the type of crop.

However, the built structure was overly heavy, as it was made of galvanized iron. Furthermore, one of the prototypes had leaky gutters (built to save weight) which caused the drainage water to run down onto the lower-level plants.

The improvements proposed for the next TRL are: to use lighter and cheaper materials such as aluminum and PVC, to design a drip irrigation system integrated into the structure itself by a quick connection, to develop a thinner and narrower structure, and to test LED light strips with other mixtures of colors such as 10:1 and 19:1.

Adopting these recommendations will help to optimize the system and increase the final yield compared to the commercial and conventional hydroponic systems.

In conclusion, I deployed a modular and low-cost innovative solution suitable for small-scale farmers and citizens, an innovation that could offer new opportunities for job creation, particularly for young or small-scale growers. The results of this study provide information to be used as a reference for managers, trainers, and growers, as well as innovators.

5. Patents

The information used for the experimental analysis is protected under Spanish Law (Utility model Number: ES 1 242 949 U).

Funding: This research received no external funding.

Institutional Review Board Statement: Not applicable.

Informed Consent Statement: Not applicable.

Data Availability Statement: The information used for the experimental analysis is protected under Spanish Law (Utility model Number: ES 1 242 949 U).

Acknowledgments: The author acknowledges the support given by the teachers, Felipe Pérez and Ángela Angulo, for the implementation of the experiment, and the university spin-off, Bo True Activities, for providing the vertical modules.

Conflicts of Interest: The author declares no conflict of interest.

References

1. United Nations, Department of Social Affairs. World Population Prospects. Available online: https://www.un.org/development/desa/en/news/population/world-population-prospects-2019.html (accessed on 15 June 2021).
2. FAO. *Regional Overview of Food Security and Nutrition in Europe and Central Asia 2019. Structural Transformations of Agriculture for Improved Food Security, Nutrition and Environment*; FAO: Budapest, Hungary, 2019; ISBN 978-92-5-132000-6.
3. European Environment Agency. Soil, Land and Climate Change. Available online: https://www.eea.europa.eu/signals/signals-2019-content-list/articles/soil-land-and-climate-change (accessed on 11 June 2021).
4. FAO Plant Production and Protection Division (AGP). Implementing the Save and Grow Approach. Building the Basis for Sustainable and Climate Resilient Cropping Systems. Available online: http://www.fao.org/sustainable-agricultural-mechanization/strategies/save-and-grow/en/ (accessed on 15 June 2021).
5. European Environment Agency. Available online: https://www.eea.europa.eu/data-and-maps/indicators/use-of-freshwater-resources-3/assessment-4 (accessed on 15 June 2021).
6. Khokhar, T. Chart: Globally, 70% of Freshwater Is Used for Agriculture. 2017. Available online: https://blogs.worldbank.org/opendata/chart-globally-70-freshwater-used-agriculture (accessed on 7 March 2021).
7. EUROSTAT. Available online: https://ec.europa.eu/eurostat/statistics-explained/index.php?title=Agri-environmental_indicator_-_mineral_fertiliser_consumption#Analysis_at_EU_level (accessed on 11 June 2021).
8. EUROSTAT. Available online: https://ec.europa.eu/eurostat/statistics-explained/index.php?title=Agri-environmental_indicator_-_consumption_of_pesticides#Analysis_at_EU_and_country_level (accessed on 11 June 2021).
9. Avgoustaki, D.D.; Xydis, G. Indoor vertical farming in the urban nexus context: Business growth and resource savings. *Sustainability* **2020**, *12*, 1965. [CrossRef]
10. Stein, E.W. The transformative environmental effects large-scale indoor farming may have on air, water and soil. *Air Soil Water Res.* **2021**, *14*, 1–8. [CrossRef]
11. Kalantari, F.; Tahir, O.M.; Lahijani, A.; Kalantari, S. A review of vertical farming technology: A guide for implementation of building integrated agriculture in cities. *Adv. Eng. Forum* **2017**, *24*, 76–91. [CrossRef]
12. De Anda, J.; Shear, H. Potential of Vertical Hydroponic Agriculture in Mexico. *Sustainability* **2017**, *9*, 140. [CrossRef]
13. Butturini, M.; Marcelis, L.F.M. Vertical Farming in Europe: Present Status and Outlook. In *Plant Factory—An Indoor Vertical Farming System for Efficient Quality Food Production*, 2nd ed.; Kozai, T., Niu, G., Takagaki, M., Eds.; Academic Press—Elsevier: Cambridge, MA, USA, 2020; pp. 77–91. [CrossRef]
14. Melkonyan, A.; Gruchmann, T.; Lohmar, F.; Kamath, V.; Spinler, S. Sustainability assessment of last-mile logistics and distribution strategies: The case of local food networks. *Int. J. Prod. Econ.* **2020**, *228*, 107746. [CrossRef]
15. Sitaker, M.; Kolodinsky, J.; Wang, W.; Chase, L.C.; Kim, J.V.S.; Smith, D.; Estrin, H.; Vlaanderen, Z.V.; Greco, L. Evaluation of Farm Fresh Food Boxes: A Hybrid Alternative Food Network Market Innovation. *Sustainability* **2020**, *12*, 10406. [CrossRef]
16. Jia, S. Local food campaign in a globalization context: A systematic review. *Sustainability* **2021**, *13*, 7487. [CrossRef]
17. Zhang, H.; Asutosh, A.; Hu, W. Implementing Vertical farming at university scale to promote sustainable communities: A feasibility analysis. *Sustainability* **2018**, *10*, 4429. [CrossRef]
18. Al-Kodmany, K. The vertical farm: A review of developments and implications for the vertical city. *Buildings* **2018**, *8*, 24. [CrossRef]
19. European Commission. Young Farmers in the EU—Structural and Economic Characteristics. Available online: https://ec.europa.eu/info/sites/default/files/food-farming-fisheries/farming/documents/agri-farm-economics-brief-15_en.pdf (accessed on 27 February 2021).
20. Kalantari, F.; Mohd, O.; Akbari, R.; Fatemi, E. Opportunities and challenges in sustainability of vertical farming: A review. *J. Landsc. Ecol.* **2017**, *11*, 35–60. [CrossRef]
21. Resh, H.M. *Hydroponic Food Production: A Definitive Guidebook for the Advanced Home Gardener and the Commercial Hydroponic Grower*; CRC Press: Boca Raton, FL, USA, 2016; ISBN 1439878676.
22. Baras, T. *DIY Hydroponic Gardens. How to Design and Build an Inexpensive System for Growing Plants in Water*; Quarto Publishing Group: Minneapolis, MN, USA, 2018; ISBN 9780760357590.
23. We Are Transforming Agriculture. Available online: https://aerofarms.com/ (accessed on 16 July 2021).
24. Zero Pesticides. Just Picked. Craveable Flavor. Available online: https://www.plenty.ag/ (accessed on 16 July 2021).
25. What Grows Here Changes Everything. Available online: https://boweryfarming.com/ (accessed on 16 July 2021).
26. Rius-Ruiz, F.X.; Andrade, F.J.; Riu, J.; Rius, F.X. Computer-operated analytical platform for the determination of nutrients in hydroponic systems. *Food Chem.* **2014**, *147*, 92–97. [CrossRef] [PubMed]
27. Ibayashi, H.; Kaneda, Y.; Imahara, J.; Oishi, N.; Kuroda, M.; Mineno, H. A reliable wireless control system for tomato hydroponics. *Sensors* **2016**, *16*, 644. [CrossRef] [PubMed]
28. Montoya, A.; Obando, F.; Morales, J.; Vargas, G. Automatic aeroponic irrigation system based on arduino's platform. *J. Phys. Conf. Ser.* **2017**, *850*, 012003. [CrossRef]
29. Eridani, D.; Wardhani, O.; Widianto, E.D. Designing and implementing the arduino-based nutrition feeding automation system of a prototype scaled Nutrient Film Technique (NFT) hydroponics using Total Disolved Solids (TDS) sensor. In Proceedings of the 2017 4th International Conference on Information Technology, Computer and Electrical Engineering ICITACEE 2017, Semarang, Indonesia, 18–19 October 2017; pp. 170–175.

30. Ruengittinun, S.; Phongsamsuan, S.; Sureeratanakorn, P. Applied Internet of Thing for Smart Hydroponic Farming Ecosystem (HFE). In Proceedings of the Ubi-Media 2017: The 10th International Conference on Ubi-media Computing and Workshops, Pattaya, Thailand, 1–4 August 2017; pp. 1–4.
31. Palande, V.; Zaheer, A.; George, K. Fully automated hydroponic system for indoor plant growth. *Procedia Comput. Sci.* **2018**, *129*, 482–488. [CrossRef]
32. Tagle, S.; Benoza, H.; Pena, R.; Oblea, A. Development of an indoor hydroponic tower for urban farming. In Proceedings of the 6th DLSU International Conference on Innovation and Technology Fair, Manila, Philippines, 22–23 November 2018; pp. 1–7.
33. Marques, G.; Aleixo, D.; Pitarma, R. Enhanced hydroponic agriculture environmental monitoring: An internet of things approach. In Proceedings of the International Conference on Computational Science, Faro, Portugal, 12–14 June 2019; pp. 658–669.
34. World Intellectual Property. Utility Models. Available online: https://www.wipo.int/patents/en/topics/utility_models.html (accessed on 20 September 2021).
35. Martínez, R.; Héctor, A.L. Double Vertical Hanging Unit, in the Form of Pouches, for Hydroponic Plant Cultivation, Panel of Double Vertical Hanging Unit, in Pouch Form, and Greenhouse Structure. EP 2 904 894 A1, 4 October 2013.
36. DeMitchell, M.; Tarzian, M. Vertical Hydroponic System. U.S. Patent 2011/0067301 A1, 24 March 2011.
37. Lin, P. Vertical Planting Apparatus. U.S. Patent 2012/0066972 A1, 31 January 2011.
38. Bryan, M. Hydroponic Plant Cultivating Apparatus. U.S. Patent 7,005,282 B2, 6 June 2006.
39. Borrero, J.D.; Fernández, G. Sistema Modular para Cultivo Hidropónico o Aeropónico. Utility Model No. ES 1 242 949 U, 6 March 2020.
40. Barta, D.J.; Tibbitts, T.W.; Bula, R.J.; Morrow, R.C. Evaluation of Light Emitting Diode Characteristics for a Space-based Plant Irradiation Source. *Adv Space Res.* **1992**, *12*, 141–149. [CrossRef]
41. Tennessen, D.J.; Singsaas, E.L.; Sharkey, T.D. Light-emitting diodes as a light source for photosynthesis research. *Photosynth. Res.* **1994**, *39*, 85–92. [CrossRef]
42. Krames, M.R.; Shchekin, O.B.; Mueller-Mach, R.; Mueller, G.O.; Zhou, L.; Harbers, G.; George Craford, M. Status and future of high-power light-emitting diodes for solid-state lighting. *J. Disp. Technol.* **2007**, *3*, 160–175. [CrossRef]
43. Trotter, M.; Whitehead, D.; Pinkney, J. The photochemical reflectance index as a measure of photosynthetic light use efficiency for plants with varying foliar nitrogen contents. *Int. J. Remote Sens.* **2002**, *23*, 1207–1212. [CrossRef]
44. Naznin, M.T.; Lefsrud, M.; Gravel, V.; Hao, X. Using different ratios of red and blue LEDs to improve the growth of strawberry plants. *Acta Hortic.* **2016**, *1134*, 125–130. [CrossRef]
45. Hanenberg, M.A.A.; Janse, J.; Verkerke, W. LED light to improve strawberry flavour, quality and production. *Acta Hortic.* **2016**, *1137*, 207–212. [CrossRef]
46. Gautier, H.; Massot, C.; Stevens, R.; Serino, S.; Genard, M. Regulation of tomato fruit ascorbate content is more highly dependent on fruit irradiance than leaf irradiance. *Ann. Bot.* **2008**, *103*, 495–504. [CrossRef] [PubMed]
47. Mochizuki, Y.; Sekiguchi, S.; Horiuchi, N.; Aung, T.; Ogiwara, I. photosynthetic characteristics of individual strawberry (*Fragaria × ananassa Duch.*) leaves under short-distance lightning with blue, green, and red LED lights. *HortScience* **2019**, *54*, 452–458. [CrossRef]
48. EPDATA. Available online: https://www.epdata.es/ (accessed on 16 July 2021).
49. Mankins, J.C. Technology readiness assessments: A retrospective. *Acta Astronaut.* **2009**, *65*, 1216–1223. [CrossRef]
50. Héder, M. From NASA to EU: The Evolution of the TRL Scale in Public Sector Innovation. *Innov. J.* **2017**, *22*, 1–23.
51. U.S. Department of Energy. *Department of Energy Technology Readiness Assessment Guide*; U.S. Department of Energy: Washington, DC, USA, 2009.
52. European Space Agency. *Technology Readiness Levels Handbook for Space Applications*; European Space Agency: Paris, France, 2008; TEC-SHS/5551/MG/ap.
53. ISO. Available online: https://www.iso.org/standard/56064.html (accessed on 16 July 2021).
54. European Commission. Technology Readiness Levels (TRL). Extract from Part 19—Commission Decision C(2014)4995. Available online: https://ec.europa.eu (accessed on 16 July 2021).
55. Sauser, B.; Verma, D.; Ramirez-Marquez, J.; Gove, R. A Systems Approach to Expanding the Technology Readiness Level within Defense Acquisition. *Int. J. Def. Acquis. Manag.* **2008**, *1*, 39–58.
56. Zipgrow. 2021. Available online: https://zipgrow.com/commercial-grower-survey/ (accessed on 16 July 2021).

 MDPI

Article

The Conceptual Synthesis and Development of a Multifunctional Lawnmower

Chun Quan Kang, Poh Kiat Ng * and Kia Wai Liew

Faculty of Engineering and Technology, Multimedia University, Jalan Ayer Keroh Lama, Bukit Beruang, Melaka 75450, Malaysia; kangchunquan@gmail.com (C.Q.K.); kwliew@mmu.edu.my (K.W.L.)
* Correspondence: pkng@mmu.edu.my

Abstract: This study aims to develop a novel, original and multifunctional lawnmower through reviews of patent literature, research literature and variants of existing lawnmowers. After a detailed conceptualisation process, Autodesk Inventor (version 2019) is used for the finalised design drawing and stress simulation. The prototype is fabricated and tested through several experiments for usability validation which included tests on sound intensity level, cutting ability, polishing performance and battery power durability. Using Minitab 19, the data for the sound intensity and cutting ability experiments are analysed with two-sample *t*-tests. The data for the polishing performance and battery power durability experiments are analysed through observations, mean comparisons, and manual calculations. Significant differences are found to exist between the tested and control parameters in the context of each experiment, with the outcomes supporting the usability and performance of the present study's multifunctional lawnmower. This study showed that the prototype has the potential to solve not only some of the problems in conventional lawnmowers but also a few limitations in existing robotic lawnmowers. The outcome of this study intends to benefit society by advancing innovation in lawn maintenance and improving quality of life.

Keywords: conceptual synthesis; design; lawnmower; multifunctionality; usability

Citation: Kang, C.Q.; Ng, P.K.; Liew, K.W. The Conceptual Synthesis and Development of a Multifunctional Lawnmower. *Inventions* **2021**, *6*, 38. https://doi.org/10.3390/inventions6020038

Academic Editors: Francisco Manzano Agugliaro and Esther Salmerón-Manzano

Received: 21 April 2021
Accepted: 21 May 2021
Published: 28 May 2021

1. Introduction

A lawnmower is a machine that people use to cut grass. It uses specific rotating blades to cut grass to a certain height. Conventional lawnmowers, which can easily be found in hardware stores, include the gasoline or petrol-powered rotary push mower and a "ride on" mower. The similarity between these two types of lawnmowers include the fact that they both require manual operations and petrol as their main power source [1]. This similarity is also considered a weakness because operating these lawnmowers consumes a considerable amount of time and manpower, especially when it comes to mowing a large field such as a golf course or sports ground. According to a lawnmower manufacturer known as Encore, mowing time and productivity can be calculated with the assumption that the lawn is flat and with no landscaping obstacles. For example, approximately 60 min is required to cut only about 4047 m^2 (1 acre) of turf for a 0.9 m (36 inch) walk behind a mower operating at 1.48 m/s (3.3 mph) [2].

Furthermore, the power source used by conventional lawnmowers is neither renewable and nor environmentally friendly. Gasoline-powered lawn and garden equipment (GLGE) such as string trimmers, stump grinders and tractors often account for elevated degrees of localised emissions, including carbon dioxide (CO_2) and criteria pollutants. According to a research carried out in the US, all non-road sources including lawnmowers are responsible for about 242 million tonnes of pollutants every year, accounting for 17% of all VOC emissions, 12% of NO_x emissions, 29% of CO emissions, 4% of CO_2 emissions, 2% of PM10 emissions, and 5% of PM2.5 emissions [3].

There are also other minor concerns about conventional lawnmowers which affect the well-being of humans such as sound pollution and safety issues. Typical gas-powered mowers can create significant noise pollutions greater than 85 decibels [4]. The sound of a mower can be heard from a quarter mile away or more. Both the World Health Organisation (WHO) and Environmental Protection Agency (EPA) recommend that people limit their total exposure to noises to 5 min per day for very loud gas mowers, around 15 min a day for averagely loud mowers and 45 min a day for quieter mowers. Any prolonged period of exposure could cause hearing impairment or loss if hearing protection is not used [5].

In terms of safety, it was found that 934,394 lawnmower injuries were treated in US Emergency Departments from 2005 to 2015, with an average of 84,944 injuries annually [6]. Other reports found that around 212,258 children were treated in the US Emergency Departments for lawnmower-related injuries between 1990 and 2014 [7]. There is often more than one type of injury caused by conventional lawnmowers, which can include cuts, burns and fractures.

Problem Statement

One of the actions taken by some experts in solving the aforementioned problems is to design and produce robotic lawnmowers. A robotic lawnmower is an autonomous robot used to cut grass. Other similar products in the market are robot vacuums and robot mops. These types of working robots usually come with similar features such as the ability to work within a specified perimeter, obstacle avoidance and safety features. Consumers often substantially emphasise on basic service robots. Due to an increased demand for these robots, the related research in this area is growing.

According to a news report published by Allied Market Research, the worldwide autonomous lawnmower market size was reported as $538 million in 2017 and is forecasted to touch $1437 million by 2025, registering a compound annual growth rate of 12.9% from 2018 to 2025 [8]. This steady advancement in market growth is one of the reasons why the robotic lawnmower market appears to grow rapidly in recent years.

Nonetheless, these solutions are still bounded by certain limitations such as high cost (the price is not proportional to the function) and high level of dependency on non-renewable resources [9]. Another issue includes the lack of alternate features in robotic lawnmowers. Multifunctionality is important as less useful equipment eventually end up being eliminated over time, while the more universal and multifunctional ones remain in use [10].

However, there has yet to be a study that systematically conceptualises and develops a multifunctional lawnmower that is capable of addressing the above-mentioned limitations in robotic lawnmowers. Hence, this study aims to conceptually synthesise and develop a multifunctional lawnmower.

It is important to note that the scope of this study covers the resolution of limitations in smart lawnmowers which comprise high cost, heavy reliance on non-renewable resources and the inadequacy of multifunctional characteristics. This study does not cover other aspects of smart lawnmower designs such as the cutting blade, micro-controller, and movement algorithm for the robot. Therefore, aspects that are beyond the scope of this study will be adopted from existing smart lawnmowers through a benchmarking process.

2. Literature Review

2.1. Locomotion Methods for Ground Robots

Locomotion in robots includes a certain energy source (such as electric, nuclear, or geothermal power) conversion into mechanical power for vehicular movement. For a robotic machine to navigate through different geographical topographies, it is necessary for a platform to have multi-terrain manoeuvring capabilities [11]. Several locomotion methods that are suitable for ground mobile robots include the wheeled locomotion, track belt or slip locomotion, and legged robot.

2.1.1. Wheeled Locomotion

Wheeled locomotion is commonly used by most moving devices such as automobiles and toys. This method allows for the achievement of high speeds while consuming less power and fewer active degrees of freedom. However, overcoming obstacles would normally be challenging with this method. To improve the stability of wheeled robots, the number of wheels is normally increased to four or more [12]. There are in total 3 different types of common wheels for autonomous robots [13]:

- Simple or standard wheels with two degrees of freedom, which are normally found around the centre shaft and point of contact.
- Castor wheels generally used in trolleys. These wheels can rotate around the axis and off-centre pivot point.
- Multidirectional wheels, which have three degrees of freedom with the help of rollers mounted on the outer periphery of the wheels.

2.1.2. Track Belt or Slip Locomotion

Using the track belt or slip locomotion eliminates some disadvantages from the wheeled locomotion. Wheeled robots can easily become stuck in a gap or hole that is bigger than the wheel size. The power efficiency for wheeled robots also drops while operating in loose terrain due to the increased rolling friction. The track belt locomotion which has a larger ground contact surface area is more efficient in loose terrain and has a lower risk of becoming stuck in holes and gaps [14].

The drawbacks of the track belt locomotion include slower movements and a higher energy consumption as compared to wheeled robots. However, these drawbacks can be minimised by lengthening the chains or changing their shape [15].

The steering system for tracked robots is different compared to wheeled robots. For tracked robots to change directions, the speed or direction of one tracked wheel needs to be different from the other. The SNR1 developed by the National Institute of Advanced Industrial Science and Technology (AIST) is one of such robots that uses a track belt locomotion. It has the capability to move on uneven terrains and staircases with the help of a single track along the wheels [11].

2.1.3. Legged Locomotion

Legged robots use articulated limbs such as leg mechanisms to provide locomotion. These robots move unlike the wheeled and track belt locomotion which intend to maintain continuous frictional contact with the ground. Many mechanisms for legged robots are imitated from humans and insects. For example, the Klann's mechanism is imitated from spiders [16].

These vehicles can move in irregular terrains by varying their leg configuration in order to adapt themselves to surface irregularities. The foot establishes contact with the ground in selected points according to terrain conditions. For these reasons, legs are inherently adequate systems for locomotion on irregular ground.

Legged robots tend to always require reprogramming since the pre-programmed ones would normally face challenges in interacting with the environment autonomously. Hence, a self-learning technique must be incorporated for these robots to adapt in different environments. In the late 1990s, researchers from MIT developed Genghis, a robot which has a reinforcement learning algorithm for walking [17]. Genghis was capable of learning the coordinates by the movement of its legs.

2.2. Materials for Robots

There are four groups of materials that tend to be commonly selected as the structural element for vehicle systems [18]. These four groups include wood, metal, composite materials, and plastics. Each group has their own characteristics, properties, potentials and disadvantages, and can be identified from the Ashby's material selection chart [19].

The indicator parameters of the selected material's mechanical properties include tensile strength, elastic modulus, and compressive strength.

2.3. Microcontrollers

A microcontroller is also known as a single board computer. It is a powerful microcomputer containing many on-board integrated circuits. Microcontrollers are created to perform a specified task and execute one specific application. This application can include automatic engine control systems, remote controls, power tools, toys and various types of office machines [20]. Some of the commonly used microcontrollers include the Arduino single-board microcontrollers [21] and Intel MCS-51 [22].

The Arduino is a more suitable choice for this study. It is a ready-to-use microcontroller board unlike the 8051, in which usage is time-consuming and error-prone, as it requires manual soldering for every pin. Furthermore, the Arduino allows an input voltage in the range of 6 to 20 volts, which means that a 12-volt battery could directly connect to the module without passing through any voltage regulator circuit as required by 8051 microcontrollers. However, there are also various types of Arduino boards available in the market such as the Arduino Uno, Arduino Mega, Arduino Micro and Arduino Nano. Thus, a comparison still needs to be done.

2.4. Patent Literature Review

2.4.1. Automated Lawn Cutting and Vacuum System

This invention not only includes the lawnmower but also lawn mowing accessories. The accessories refer to the mower home base and automated vacuum system. The lawn is equipped with multiple perimeter sensors placed around the perimeter and a mower home base which is placed at the corner of the lawn.

The mower home is used to store and recharge the lawnmower when it is not in use and remove the grass clippings. All tasks are performed automatically by the lawnmower, including the cutting and vacuuming of the grass, recharging, and storage and removal of grass clippings. The perimeter sensors are used to specify the working areas for the lawnmower [23].

2.4.2. Automated Lawnmower or Floor Polisher

The invention relates to an improved robotic lawnmower which uses obstacle- and grass touch-sensing for the motion control of a lawnmower through a computer interfaced feedback control. The invention uses infrared obstacle detection together with cut or uncut sensing to deliver feedback control to independent wheel drive motors when manoeuvring the lawnmower.

This invention aims to address the disadvantage of using a conventional automated lawnmower which requires a programmed lawn plot map or boundary track indicators such as perimeter sensors to work within a specified boundary. Alternatively, the automated lawnmower can also be used as a floor polisher by simply substituting the cutting blade with a polishing pad [24].

2.4.3. Multifunctional Mobile Appliance

This invention aims to perform multiple tasks which include mowing, vacuuming, mulching, polishing, weeding, painting, edging, fertilising, cultivating, raking, sanding, pressure washing or shampooing. The invention is also equipped with a work module distribution mechanism where all work modules are stored so that the mobile unit can switch them according to the respective tasks automatically. The invention is also embedded with a real-time kinematic global positioning system [25].

2.5. Journal Review

2.5.1. The "Automated Mower Robo"

A project entitled the "Automated Mower Robo" was proposed in order to solve the problems of using conventional lawnmowers powered by petrol. The project aimed to prototype an environmentally friendly electric solar grass cutter.

The design includes a solar-powered autonomous robot that communicates with the user through a smartphone's Bluetooth module. The design uses Arduino and ultrasonic sensors for obstacle detection. The prototype was tested on four different species of grass. The results in Table 1 show that the automated lawnmower is able to cut the grass to a minimum height of 80 mm [26].

Table 1. Calculated results on grass height for different types of grass before and after mowing [26].

Types of Grass	Height (mm)		
	Mean Grass Height (Before Mowing)	Mean Grass Height (After Mowing)	Expected Grass Height (After Mowing)
Elephant grass	230.0	80.0	90.0
Stubborn	234.0	84.0	90.0
Spare grass	111.0	80.0	80.0
Carpet grass	70.5	-	-

2.5.2. Multi Sensor and Multifunctional Robot with Image Mosaic

This invention was reported to possess the features of obstacle avoidance and line-following. The motion of the robot was based on clap sounds and light sensors. In this paper, the researchers reported that the line-following robot moves along the line using a certain feedback mechanism. Light sensors allow the robot to move according to the light and clap sensors allow the detection of clap sounds used to turn the robot on and off [27].

3. Methodology

The following steps are adhered to during the conceptualisation process:

- Step 1: Structural concept generation with combination chart
- Step 2: Screening and scoring of structural concepts
- Step 3: Key features scoring and combinations
- Step 4: Final concept generation
- Step 5: Material selection
- Step 6: Design drawing and stress simulation

3.1. Structural Concept Generation with Combination Chart

The basic components for a multifunctional lawnmower can be classified into two different groups, namely the mechanical and electronic components. The mechanical components include the mower shape, locomotion types and blade types. It is important to note that nylon string cutters (one of the substitutes for blades) are not included in the selection as these cutters often wear out faster than metal blades and are not suitable for smart lawnmowers that require minimal maintenance [28]. The electronic components comprise the microcontroller, battery and motor types for the blade and wheels. All the common choices for each component are listed in a combination chart as shown in Table 2.

Table 2. List of all common choices for each component.

Mechanical	Driving mechanism	• Continuous track • 4 equal size wheels • 2 big rear wheels and 1 round ball wheel at the front
	Shape	• Circular • Rectangular
	Blade type	• Large individual metal cutting blade • Razor blades attached to a spinning disc • Two swinging blades at the front of the mower
Electronic	Battery type	• Lead acid • Lithium-ion • Nickle-cadmium
	Motor type (blade)	• Servo DC motor • Brushless DC motor • Stepper motor • Brushed DC motor
	Motor type (wheels)	• Gear motor
	Arduino board types	• Arduino Uno • Arduino Mega • Arduino Micro • Arduino Nano

Different combinations of mechanical components lead to different basic structural concepts for the lawnmower. For this case, there are in total 5 different structural concepts. Table 3 shows the hand-sketched drawings of the bottom and isometric views for all the structural concepts.

Table 3. Bottom and isometric views of the structural concepts (hand sketches).

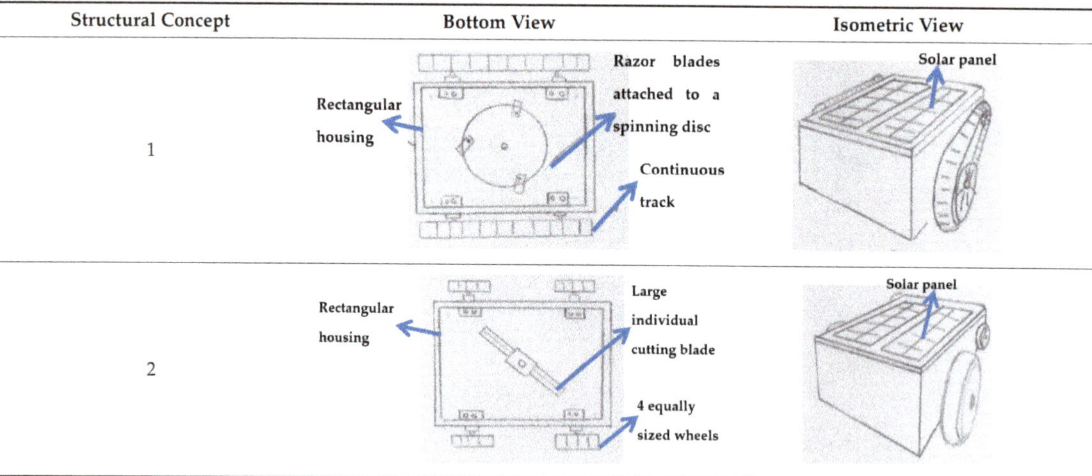

Table 3. *Cont.*

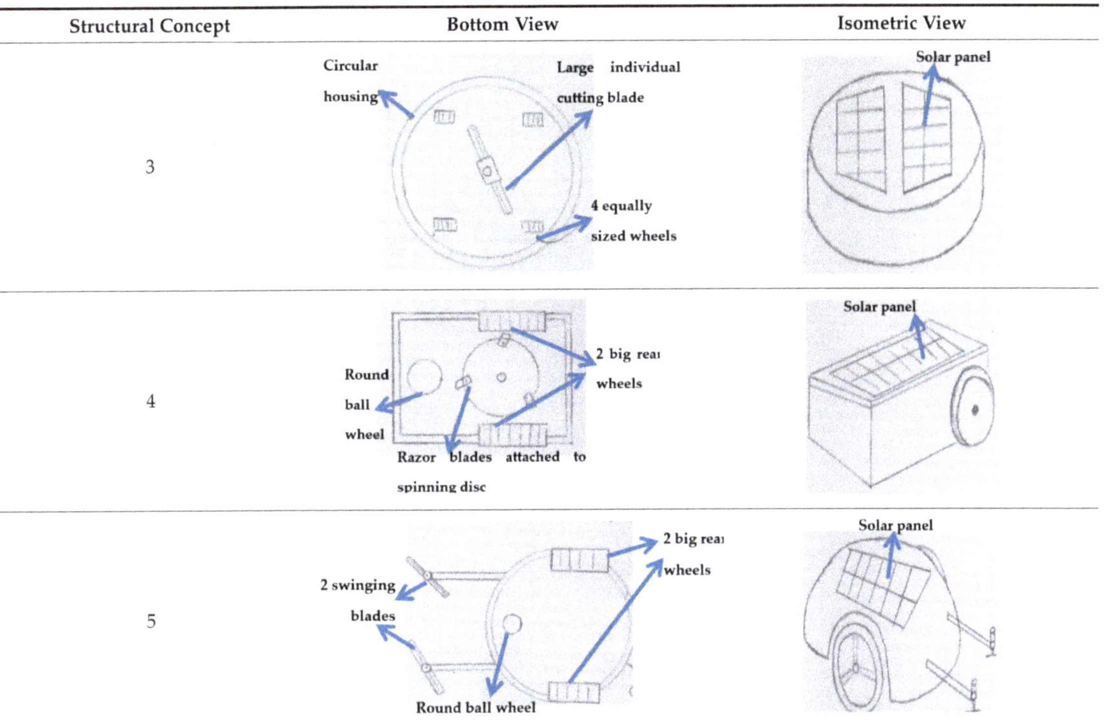

Structural concept 1 includes a rectangular-shaped robot with a continuous track driving mechanism and razor blades attached to a spinning disc. Structural concept 2 uses a rectangular-shaped robot with four equally sized wheels and a large individual cutting blade. Structural concept 3 combines a circular-shaped robot with four equally sized wheels and a large individual cutting blade. Structural concept 4 integrates a rectangular-shaped robot with two big rear wheels, one round ball wheel at the front, and razor blades attached to a spinning disc. Lastly, structural concept 5 combines a circular-shaped robot with two big rear wheels, one round ball wheel at the front, and two swinging blades at the front.

3.2. Screening and Scoring of Structural Concepts

During the screening process shown in Table 4, it is found that only structural concepts 1, 2, and 4 are selected for the subsequent scoring process since the net scores of these concepts are found to be higher than the net score for structural concept 3 (the concept of reference).

It is important to note that the scoring and ranking processes are done by the main author of this study with some advice from his co-authors. The weightage and ratings are proposed by the main author based on his specific experience and knowledge of various lawnmower designs. The main author is also in the forefront of the design work and has a good grasp of the prototyping requirements, cost and ergonomics aspects involved in this study. Therefore, in reference to the main author's superior design sense in the specific area of lawnmower designs, the co-authors of this study concurred to the ratings and rankings provided by the main author.

The rating system used for the scoring process ranged from 1 to 5, and the description for each score is defined as such:

- 1: much worse than expected
- 2: worse than expected
- 3: matches expectations
- 4: exceeds expectations
- 5: greatly exceeds expectations

For the scoring process, each selection criterion is given a weight (W). To determine the weighted score (WS) of each criterion, the rating (R) is multiplied by the weight (WS = R × W). The total weighted score for each structural concept is compared, and the structural concept with the highest score is prioritised for further development.

Table 4. Screening of basic structural concepts.

Selection Criteria	Basic Structural Concept				
	1	2	3 *	4	5
Ease of manufacturing	0	+	0	+	–
Cutting width	0	0	0	0	+
Cost	–	+	0	+	–
Stability	+	+	0	+	–
Edging possibility	–	0	0	–	+
Ease of maintenance	+	0	0	0	0
Safety	+	0	0	+	–
Sum of +	3	3	0	4	2
Sum of 0	2	4	7	2	1
Sum of –	2	0	0	1	4
Net score	1	3	0	3	−2
Rank	2	1	3	1	4
Continue?	YES	YES	NO	YES	NO

Note: * Benchmarked concept.

The weight of each selection criterion depends on the constraints in the project. The major constraints include the project time and budgetary limitations. As it is important to ensure that the project is completed within these constraints, criteria such as ease of maintenance and ease of manufacturing are given high weights (30% each). It is also highly desired that the multifunctional lawnmower performs optimally with regard to functionality which accounts for edging possibilities, stability and multifunctionality. Therefore, the weight for functionality is set at 40%. During the scoring process, structural concept 4 obtained the highest score and is hence selected as the lawnmower's basic structure (refer to Table 5).

Since the electronic parts are not emphasised in the visualisation of the lawnmower's basic structure, their screening process is done individually. The screening processes for the battery, blade motor and Arduino board types are shown in Tables 6–8. In the end, the lead acid battery, brushed DC motor and Arduino Uno are selected. For the most part, the lead acid battery and brushed DC motor are selected over the other components due to lower cost, while the Arduino Uno is selected because it is easily accessible and less complex as contrasted to the other types of Arduino boards.

Table 5. Scoring of basic structural concepts.

Selection Criteria	W (%)	Lawn Mower Base Structure					
		SC 1		SC 2		SC 4	
		R	WS (%)	R	WS (%)	R	WS (%)
Functionality	40						
Usage in different lawns	10	4	40	3	30	4	40
Zero turning radius	5	5	25	4	20	5	25
With high stability	10	4	40	2	20	3	30
Wide cutting width	5	4	20	4	20	3	15
Edging possibility	10	2	20	3	30	3	30
Ease of maintenance	30						
Easily replaceable cutting blade	15	4	60	3	45	4	60
Simple fastening methods	15	3	45	3	45	3	45
Ease of manufacturing	30						
Low-cost material	10	2	20	3	30	3	30
Low complexity of parts	10	3	30	4	40	3	30
Low number of assembly steps	10	2	20	3	30	4	40
Total score		320		310		345	
Rank		2		3		1	
Continue?		NO		NO		YES	

Notes: W—Weight; SC—Structural concept; R—Rating; WS—Weighted score.

Table 6. Screening of battery types.

Selection Criteria	Battery Types		
	Lead Acid	Nickle-Cadmium *	Lithium-Ion
Lifecycle	−	0	−
Rechargeability	0	0	0
Safety	0	0	0
Cost	+	0	−
Power density	0	0	+
Self−discharge ability	+	0	0
Maintenance requirement (discharge)	+	0	+
Charging time	0	0	+
Sum of +	3	0	3
Sum of 0	4	8	3
Sum of −	1	0	2
Net score	2	0	1
Rank	1	3	2
Continue?	YES	NO	NO

Note: * Benchmarked battery.

The design process is extended to the detailed selection of each component. These components include the motors and battery. For the gear motor (drive motor), the maximum torque without slippage from the rear wheel is calculated to estimate the approximate torque required by the gear motor. This maximum torque without slippage is calculated using the basic equation, $T = F \times r$, where F refers to tangential force (Newton) and r refers to the radius of the wheel (metre).

Figure 1 shows the free-body diagram for the driving wheel. To perform the preceding calculation, the tangential force (F), normal force (N) and friction coefficient (μ) need to be known first. The N is calculated as 23.125 N (obtained from static stress analysis at preliminary design stage), and the friction coefficient for the contact between the wheel and dry grass is between 0.5 to 0.8.

Table 7. Screening of blade motor types.

Selection Criteria	Motor Type (Blade)			
	Stepper Motor *	Brushed DC Motor	Brushless DC Motor	Servo DC Motor
Cost	0	0	–	–
Noise	0	+	+	+
Speed	0	+	0	+
Weight	0	0	0	0
Efficiency	0	0	+	0
Sum of +	0	2	2	2
Sum of 0	5	3	2	2
Sum of	0	0	1	1
Net score	0	2	1	1
Rank	3	1	2	2
Continue?	NO	YES	NO	NO

Note: * Benchmarked motor.

Table 8. Screening of Arduino board types.

Selection Criteria	Arduino Board Type			
	Arduino Uno	Arduino Mega	Arduino Micro *	Arduino Nano
Cost	0	–	0	0
Availability	+	–	0	–
Device complexity	+	–	0	0
Sum of +	2	0	0	0
Sum of 0	1	0	3	0
Sum of –	0	3	0	1
Net score	2	−3	0	−1
Rank	1	4	2	3
Continue?	YES	NO	NO	NO

Note: * Benchmarked Arduino board.

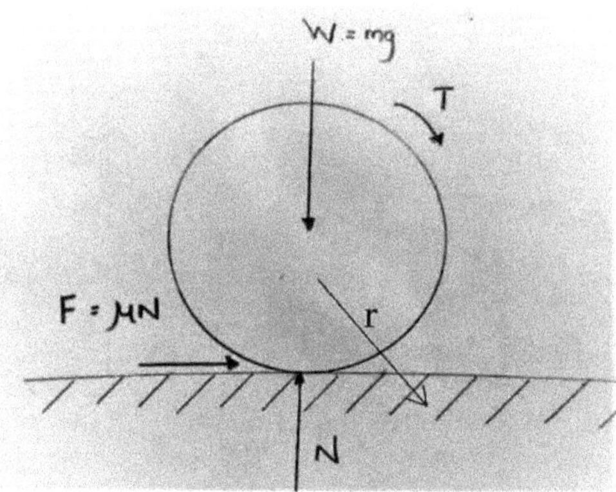

Figure 1. Free-body diagram for the driving wheel.

Therefore, the maximum torque without slippage can be calculated as $T = \mu Nr =$ (0.8) (23.125) (0.075) = 1.388 Nm. Since the friction coefficient used in the equation is an estimated value, a gear motor with a torque that is close to the calculated maximum torque value (1.388 Nm ± 10%) is acceptable. The selected gear motor is a 12 V 1 A, 24 rpm and 1.47 Nm DC motor.

The other type of motor required by the blades is the high-speed brushed DC motor. The motor requires a minimum rotational speed of 3000 rpm in order to create enough force and speed for the blades to cut grass. It is preferable for this motor to possess the same rated voltage (12 V) as the gear motor as this condition standardises the design and facilitates the selection of the other parts (battery and motor controller). The selected high-speed brushed DC motor is a 12 V 1.3 A, 10,000 rpm and 0.032 Nm DC motor.

Since the rated voltage for both motors is 12 V, a 12 V battery is selected. However, the ampere-hour (Ah) for the battery still needs to be determined according to the requirements. The battery should be a lead-acid battery at an acceptable size (able to fit the smart lawnmower), with a minimum operating time of 2 h and good availability. The minimum Ah value of the battery can be calculated as the total current × minimum operating hours.

The total current for the three motors is calculated as 3.3 A (1 A + 1 A + 1.3 A). Hence, the minimum ampere-hour for the battery is 6.6 Ah, whereby Ah = (3.3) (2) = 6.6 Ah. Hence, a 12 V battery with 7.2 Ah is selected for this study.

3.3. Key Features Scoring and Combinations

Based on the benchmarking done on existing products, patents and journals, a total of 18 different features are extracted (refer to Table 9). This list also included some self-generated ideas. Since it is not possible to accommodate every single feature into one device, a scoring process is also done on the listed features.

Table 9. List of features from different sources.

Sources	Product/Design	Key Features
Product Review	Robomow RX12 [29]	1. With unique edge mode (able to trim most of the grass at the edge). 2. Able to mow high grass. 3. Adjustable blade height.
	Husqvarna Automower 105 [30]	4. Works in any weather including raining days. 5. Theft-proofed with a pin code and alarm. 6. With lift sensors that stop blades when lawnmower is lifted.
	iRobot Terra [31]	7. Moves up and down in stripes (a square zig-zag path). 8. Able to cut close to the edge.
	Agrirobot	9. With 3 types of charging modes; inductive, contact-based and with ZCS active panel (solar energy). 10. Able to mow high grass.
Patent	Automated lawn cutting and vacuum system [23]	11. With automated vacuum system. 12. Includes mower home base for recharging and removal of grass clippings.
	Automated lawn mower or floor polisher [24]	13. Able to work as a floor polisher after the mower blade is substituted by a floor polishing pad.
	Multifunctional mobile appliance [25]	14. Able to perform multiple tasks which include lawn mowing, waxing, fertilising, polishing, edging and vacuuming.
Journals	Automated Mower Robo [26]	15. Charges from sunlight while mowing the lawn.
	Multi sensor and multifunctional robot with image mosaic [27]	16. With a sound sensor (clap sounds to turn the mower on and off).
Self-generated ideas	-	17. With water sprinkler system for watering grass. 18. Able to work automatically or by remote control.

The scoring process for the listed features is shown in Table 10 with a rating system that is similar to previous scoring process. For this scoring process, probability weights are used for each selection criteria. After the scoring process, the features with a total score of less than 3 are omitted from the selection process. Table 11 shows the 11 features with scores that are equal or above 3. Since some of the products or designs share the same feature, the total number of key features is reduced as shown in Table 10 (For example feature 1 and feature 8).

Table 10. Key features scoring table.

Key Features		Selection Criteria with Probability Weights (Total Weight = 1.0)				Score	Rank	Continue?
		Cost (0.3)	Simplicity (0.2)	Feasibility (0.3)	Suitability (0.2)			
1, 8	R	2	3	4	4	3.2	6	YES
	WS	0.6	0.6	1.2	0.8			
2, 10	R	2	4	4	4	3.4	4	YES
	WS	0.6	0.8	1.2	0.8			
3	R	3	2	4	4	3.3	5	YES
	WS	0.9	0.4	1.2	0.8			
4	R	3	4	5	4	4.0	1	YES
	WS	0.9	0.8	1.5	0.8			
5	R	1	1	2	2	1.5	11	NO
	WS	0.3	0.2	0.6	0.4			
6	R	2	2	4	4	3.0	7	YES
	WS	0.6	0.4	1.2	0.8			
7	R	4	2	2	4	3.0	7	YES
	WS	1.2	0.4	0.6	0.8			
9	R	1	1	2	3	1.7	10	NO
	WS	0.3	0.2	0.6	0.6			
11	R	3	3	4	4	3.5	3	YES
	WS	0.9	0.6	1.2	0.8			
12	R	1	3	3	3	2.4	8	NO
	WS	0.3	0.6	0.9	0.6			
13	R	4	5	4	2	3.8	2	YES
	WS	1.2	1.0	1.2	0.4			
14	R	1	1	1	2	1.2	12	NO
	WS	0.3	0.2	0.3	0.4			
15	R	2	3	4	4	3.2	6	YES
	WS	0.6	0.6	1.2	0.8			
16	R	2	2	2	1	1.8	9	NO
	WS	0.6	0.4	0.6	0.2			
17	R	3	3	3	3	3.0	7	YES
	WS	0.9	0.6	0.9	0.6			
18	R	4	2	4	3	3.4	4	YES
	WS	1.2	0.4	1.2	0.6			

Notes: R—Rating; WS—Weighted score.

Table 11. List of selected key features with their respective total score.

Selected Lawnmower Features	Total Score
1. Able to cut close to the edge.	3.2
2. Able to mow high grass.	3.4
3. Adjustable blade height.	3.3
4. Works in any weather including raining days.	4.0
5. With lift sensors that stop blades when lawnmower is lifted.	3.0
6. Moves up and down in stripes (a square zig-zag path).	3.0
7. With automated vacuum system.	3.5
8. Able to work as a floor polisher.	3.8
9. Charges from sunlight while mowing the lawn.	3.2
10. With water sprinkler system for watering grass.	3.0
11. Able to work automatically or by remote control.	3.4

These 11 selected features are further divided into five different groups with some constraints applied. First, each feature is allowed to only appear twice in different combinations. Secondly, the total score for each combination is to be within a range of 16 to 18. The purpose of these constraints is to ensure that the five different combinations would be able to compete with one another in a fair situation.

To prolong the servicing time of the lawnmower and reduce the usage of non-renewable energy when charging the lawnmower, feature 9 (Charges from sunlight while mowing the lawn) from Table 11 is selected as a pre-requisite feature and included in every combination. Table 12 shows the 5 different feature combinations along with their respective total scores. These combinations are prioritised for the final concept generation stage since their scores are within the prescribed range.

Table 12. List of combined features with their respective total scores.

No.	Combination	Total Combination Score
1	1 + 2 + 3 + 7 + 9	16.6
2	2 + 4 + 5 + 6 + 9	16.6
3	1 + 8 + 9 + 10 + 11	16.6
4	4 + 6 + 7 + 9 + 10	16.7
5	3 + 5 + 8 + 9 + 11	16.7

3.4. Final Concept Generation

The five different combinations identified from the preceding section are integrated into the previously selected structural concept 4 to form the final concepts used to solve the problems deliberated in this study. The following are descriptions of the final concepts.

- Final concept 1: This concept includes an automated solar-charged lawnmower with a vacuum system. It can be used to cut tall grass to a desirable height. It has an "edge mode" feature that allows it to cut and vacuum close to the lawn's edge.
- Final concept 2: This concept includes an automated solar-charged lawnmower that also works on rainy days. It can be used to cut tall grass that is within a specified path. It has lift sensors to detect whenever the lawnmower is lifted up or turned over.
- Final concept 3: This concept includes a solar-charged two-mode (automated or remote control) lawnmower or floor polisher with a water sprinkler system. It has an "edge mode" feature that allows it to cut and vacuum close to the lawn's edge.
- Final concept 4: This concept includes an automated solar-charged lawnmower with a vacuum and water sprinkler system. The lawnmower is programmed such that it can move in a specified path and can operate even during rainy days.
- Final concept 5: This concept includes a solar-charged two-mode (automated or remote control) lawnmower or floor polisher with lift sensors. It is a lawnmower that can also work as a floor polisher by just substituting the cutting blade with a polishing pad and adjusting the height of the rotating pad wherever needed.

The similar screening process is applied in order to obtain the best few final concepts. Table 13 shows the screening process for the final concepts, in which concepts 2, 3 and 5 are selected. Finally, the best final concept is determined by a scoring process, in which concept 5 is chosen as shown in Table 14.

Table 13. Screening of final concepts.

Selection Criteria	Final Concept				
	1	2	3 *	4	5
Ease of manufacturing	−	+	0	−	−
Ease of use	+	+	0	−	+
Cost	−	0	0	−	−
Stability	+	+	0	0	+
Feasibility	0	0	0	0	0
Multifunctionality	−	0	0	+	+
Ease of maintenance	0	0	0	−	−
Ease of manufacturing	−	+	0	−	−
Sum of +	2	3	0	1	3
Sum of 0	2	4	7	2	1
Sum of −	3	0	0	4	3
Net score	−1	3	0	−3	0
Rank	3	1	2	4	2
Continue?	NO	YES	YES	NO	YES

Note: * Benchmarked final concept.

Table 14. Scoring of final concepts.

Selection Criteria	W (%)	Multifunctional Lawnmower					
		FC 2		FC 3		FC 5	
		R	WS (%)	R	WS (%)	R	WS (%)
Functionality	50						
Safe to use	10	3	30	2	20	4	40
With high stability	5	4	20	2	10	3	15
Multifunctionality	10	2	20	4	40	4	40
Reduced human interaction	15	5	75	4	60	4	60
Environmentally friendly	10	4	40	4	40	4	40
Ease of use	30						
Minimal before-and-after cutting work	10	3	30	4	40	3	30
Simple fastening methods	5	4	20	3	15	4	20
Ease of maintenance	15	3	45	3	45	3	45
Ease of manufacturing	20						
Low-cost material	10	3	30	2	20	2	20
Low complexity of parts	5	3	15	4	20	5	25
Low number of assembly steps	5	3	15	3	15	3	15
Total Score		340		325		350	
Rank		2		3		1	
Continue?		NO		NO		YES	

Notes: W—Weight; FC—Final concept; R—Rating; WS—Weighted score.

The final concept selected is the solar-charged two-mode lawnmower or floor polisher with lift sensors. Figure 2 shows the concept's hand-sketched drawing.

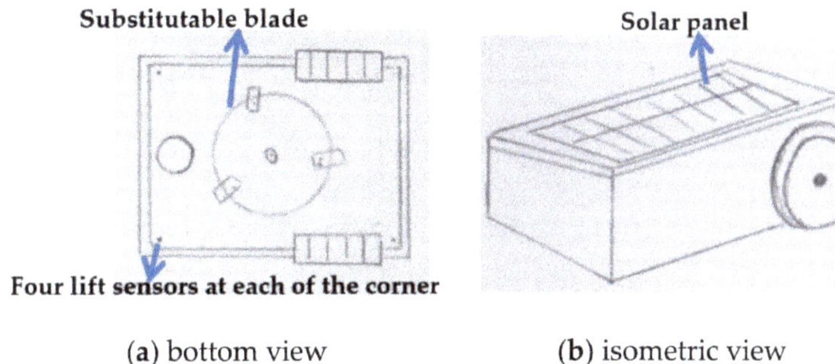

(a) bottom view (b) isometric view

Figure 2. Hand-sketched drawing of final concept.

3.5. Material Selection

The material selection is performed for the multifunctional lawnmower's body (shell). The four groups of materials that are commonly used as structural elements for vehicle systems include wood, metal, composite materials, and plastics (from Section 2.2). The following material selection procedures are applied.

- Defining design requirements and deriving material indices
- Referencing the Ashby chart
- Screening and scoring

The design requirements are tabulated in Table 15, which comprises the main functions (the function of the lawnmower's body), objectives of using the "Bubble Chart" and constraints (fixed or must-have achievements by the design).

Table 15. Design requirements for lawnmower body.

Functions	The Shell and Chassis Support the Mower's Weight and Impact Load
Objectives	1. Minimise the mass (to enhance efficiency) 2. Minimise thickness (to reduce material cost)
Constraints	1. Length is fixed 2. Must not break or crack if physically struck by accident

To achieve the design requirement, the material must be able to withstand buckling and bending loads to a certain extent with minimal density. The material indices for buckling and bending are similar and represented as $M_2 = E/\rho$, where E refers to Young's modulus, and ρ refers to density. However, there is another material index applied in the Ashby chart which is expressed as $M_1 = E$. This material index is used to specify the minimum value of the material's Young's modulus. Since wood generally has a relatively low E value compared to the other 3 groups of materials, M_1 is set at the wood region where E is around 1 GPa. The establishment of M_1 helps eliminate about half of the materials in the Ashby chart.

The second material index, M_2 identifies materials with higher strength but lower density. With the establishment of material indices M_1 and M_2, only a quarter of the materials are left to be considered in the Ashby chart. Table 16 shows the remaining 4 materials selected from the Ashby chart after screening the materials using the material indices.

Based on the screening process in Table 17, wood is selected among the 4 materials. However, there are 3 types of commonly used wood which include hardwood, softwood, and manufactured board. Further screening revealed that the manufactured board is the most suitable wood type for this study (see Table 18). Lastly, different types of manufactured boards (medium density fibre board, plywood, hardboard, and chipboard) are scored using probability weights for each selection criteria. In the end, plywood is chosen to be the shell of the lawnmower (see Table 19).

Table 16. Selected materials from Ashby chart.

Material	M_1 (GPa)	M_2 (GPa·m^3/Mg)	Comments
Wood	| grain	8–20	10–35	Good M_1; average M_2; cheap; common; reliable
CFRP	80–200	45–120	Outstanding M_1; good M_2; very expensive; aesthetic
GFRP	18–30	9–17	Good M_1; poor M_2; expensive
Steel	185–230	29–22	Outstanding M_1; good M_2; heavy; common

Table 17. Screening of materials selected from "Bubble Chart".

Selection Criteria	Materials from the Ashby Chart			
	Wood	CFRP	GFRP *	Steel
Cost	+	–	0	+
Mass	–	+	0	–
Ease of machining	0	0	0	–
Ease of handling	+	0	0	+
Ease of fastening	+	0	0	–
Good heat insulator	0	–	0	–
Sum of +	3	1	0	2
Sum of 0	2	3	6	0
Sum of –	1	2	0	4
Net score	2	−1	0	−2
Rank	1	3	2	4
Continue?	YES	NO	NO	NO

Note: * Benchmarked material.

Table 18. Screening of different wood type.

Selection Criteria	Wood Type		
	Hardwood	Softwood *	Manufactured Board
Cost	–	0	0
Mass	–	0	–
Availability	–	0	+
Ease of machining	–	0	+
Moisture resistance	+	0	–
Hardness	+	0	+
Sum of +	2	0	3
Sum of 0	0	6	1
Sum of –	4	0	2
Net score	−2	0	1
Rank	3	2	1
Continue?	NO	NO	YES

Note: * Benchmarked wood.

Table 19. Scoring of different manufactured board type.

Selection Criteria	PW (Total = 1.0)	Manufactured Board Type							
		MDFB		Plywood		Hardboard		Chipboard	
		R	WS	R	WS	R	WS	R	WS
High hardness	0.3	3	0.9	5	1.5	4	1.2	1	0.3
Ease of machining	0.2	4	0.8	3	0.6	3	0.6	2	0.4
Low material cost	0.3	4	1.2	3	0.9	3	0.9	4	1.2
Low material mass	0.2	3	0.6	3	0.6	4	0.8	4	0.8
Total score		3.5		3.6		3.5		2.7	
Rank		2		1		2		3	
Continue?		NO		YES		NO		NO	

Notes: PW—Probability weight; MDFB—Medium density fibre board; R—Rating; WS—Weighted score.

3.6. Design Drawing

The dimension of the lawnmower in this study is first estimated by benchmarking three similar products known as the Robomow RX12, Husqvarna Automower 105 and iRobot Terra. Table 20 shows the estimated size (body dimension and rear wheel diameter) and mass for the robotic lawnmower. The estimated dimensions for length, width and height are 530 mm, 400 mm, and 240 mm, respectively. These values are referenced as input values for the design modelling using the Autodesk Inventor (version 2019).

Table 20. Estimation for the size and mass of the multifunctional lawnmower.

Product	Length (mm)	Width (mm)	Height (mm)	Mass (kg)	Rear Wheel Diameter (mm)
1	508.00	419.10	254.00	7.26	≈200.00
2	550.00	390.00	250.00	6.70	≈200.00
3	530.00	405.00	202.00	8.50	≈200.00
Mean	529.33	404.70	235.33	7.49	200.00
Estimation	530.00	400.00	240.00	8.00	200.00

Notes: 1—Robomow RX12; 2—Husqvarna Automower 105; 3—iRobot Terra.

After all the components are finalised and confirmed, the assembly of the prototype commenced with parts such as the wheels (front and back), cover, blades, motors and sensors. Table 21 shows the isometric view and bottom view of the robotic lawnmower's shell and assembly drawings.

Table 21. Isometric and bottom view of the lawnmower (shell and assembly).

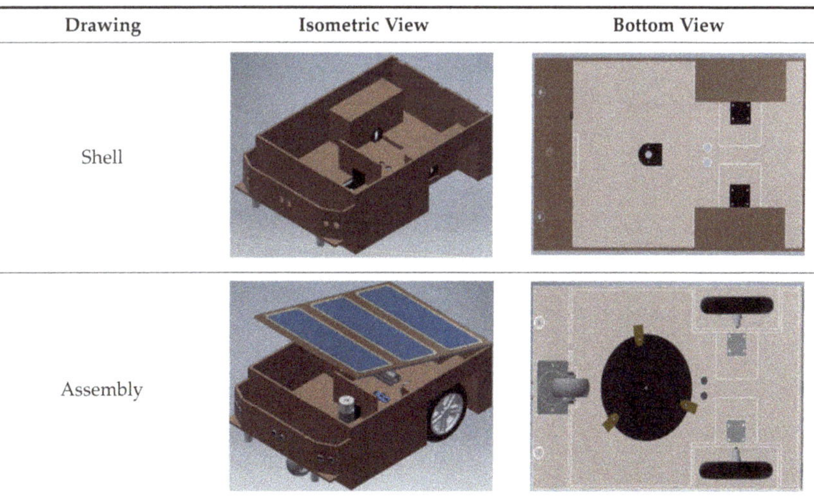

Drawing	Isometric View	Bottom View
Shell		
Assembly		

3.7. Stress Analysis

For the stress analysis, the reaction force on the body needs to be identified using static force analysis. Figures 3 and 4 show the free-body diagram of the lawnmower and the simplified free-body diagram with known and unknown variables respectively. The static force equilibrium equation when the lawnmower is in a static equilibrium state includes the summation of moment and summation of force acting on the system equated to zero:

$$\Sigma M = 0 \quad (1)$$

$$\Sigma F = 0 \quad (2)$$

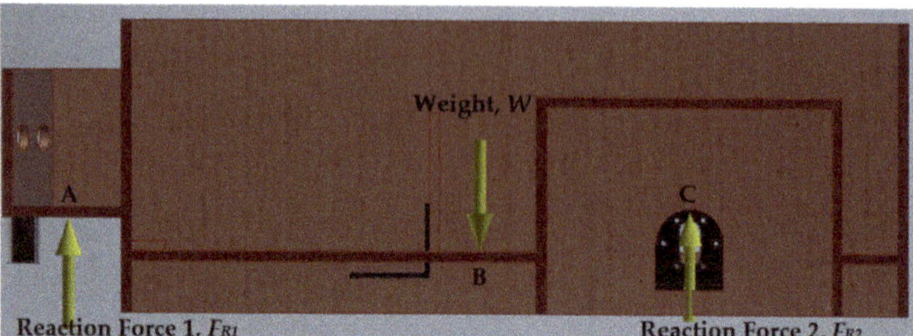

Figure 3. Free-body diagram of the lawnmower.

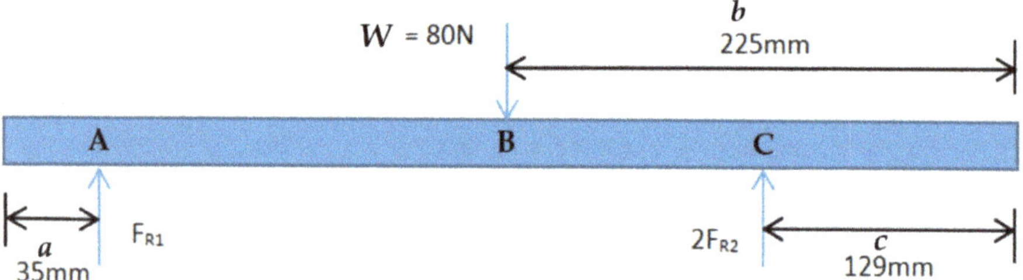

Figure 4. Simplified free-body diagram with known and unknown variables. Notes: *a*—distance from lawnmower front to point A; *b*—distance from point B to lawnmower rear; *c*—distance from point C to lawnmower rear; F_{R1}—reaction force 1; F_{R2}—reaction force 2; *W*—lawnmower weight.

The total weight, *W* of the lawnmower is estimated to be 80 N (since its mass is about 8 kg) and the following assumptions are made.

Assumptions:

1. The total weight is acting at the centre of the lawnmower.
2. Roller reaction is at the front wheel.
3. Pin reaction is at the rear wheel.

Based on the static force equilibrium equation and assumptions, the pin reaction at the front (F_{R1}) is contributed by the front wheel, while the two pin reactions (F_{R2}) are contributed by the two rear wheels of the lawnmower.

Based on Equations (3) and (4), the reaction forces F_{R1} and F_{R2} are directly proportional to the weight, *W* of the lawnmower. Other information obtained from both equations include the reaction force F_{R2} which is also directly proportional to the length denoted as *c* and inversely proportional to length denoted as *b*. Furthermore, F_{R1} can be reduced by increasing the length denoted as *b* and/or increasing the length denoted as *c*. Based on the finalised design of the lawnmower, the lengths denoted as *a*, *b* and *c* are represented as 35 mm, 225 mm, and 129 mm, respectively.

$$
\begin{gathered}
+CCW \sum M_C = 0 \\
+CCW \sum M_C = W(b-c) - F_{R1}(b) = 0 \\
F_{R1} = \frac{W(b-c)}{b} \\
+\uparrow \sum F_y = 0 \\
+\uparrow \sum F_y = F_{R1} + 2F_{R2} - W = 0 \\
F_{R2} = \frac{W - F_{R1}}{2}
\end{gathered}
\quad (3)
$$

$$F_{R2} = \frac{W - \frac{W(b-c)}{b}}{2}$$

$$F_{R2} = \frac{Wc}{2b} \tag{4}$$

After identifying all the reaction forces, these values are used for stress simulation with Autodesk Inventor (version 2019) in order to observe the stress distribution on the lawnmower in a static equilibrium state. Table 22 shows the simulation results for the von Mises stress, displacement, and safety factor. The maximum stress (0.4691 MPa) occurs at the rear wheel's mounting. On the other hand, the maximum displacement (0.009708 mm) after simulation occurs at the edge of the front surface. The minimum safety factor of the lawnmower in a static equilibrium state is 15, which is a value that is higher than 1. This outcome indicates that although a deflection of around 0.009708 mm exists at the base plate, the lawnmower is likely to still be structurally safe for use without any risks of static failure. Table 23 shows pictures of the prototype lawnmower.

Table 22. Stress simulation results for the prototype.

Stress Analysis		
Von Mises Stress (MPa)	**Displacement (mm)**	**Safety Factor**

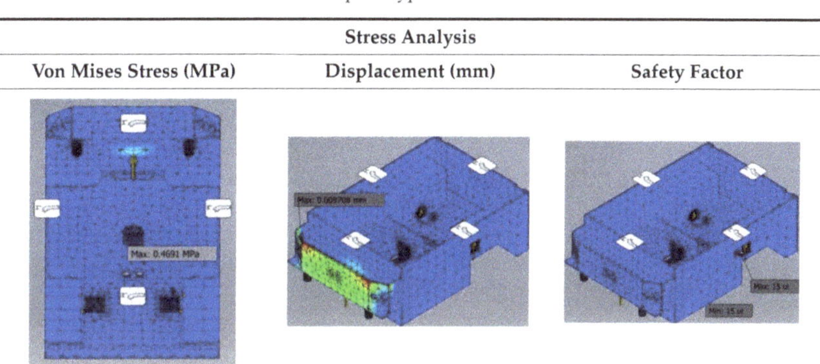

Table 23. Pictures of the prototype lawnmower.

Top-Side View	Top-Front View	Bottom View	Expanded View

4. Usability Tests

A few usability tests were conducted with the prototype multifunctional lawnmower in order to test out its effectiveness in several functions. The following tests were carried out:

- Sound intensity level experiment
- Cutting ability experiment
- Polishing performance experiment

- Battery power durability experiment

4.1. Sound Intensity Level Experiment

The objective of this experiment is to investigate whether there is a significant difference in sound intensity between a conventional lawnmower and a multifunctional lawnmower. For this experiment, the researchers were unable to compare the prototype with an existing smart lawnmower due to budgetary constraints. However, this study hypothesises that the prototype at least functions more quietly than a conventional lawnmower.

Thirty (30) sound intensity data samples (in dB) are collected from the operating noise produced by the present study's multifunctional lawnmower and a conventional lawnmower. The sound level data are measured using a sound meter. The sound meter showed that the sound produced by the conventional lawnmower is akin to that of a busy street, while the sound produced by this study's lawnmower is akin to that of a quiet library.

Further statistical analysis is carried out using Minitab (version 19). The statistical test used is the two-sample *t*-test. The null and alternative hypotheses for this test are stated as such:

H_{0a}. *There is no significant difference in the sound intensity level between a conventional lawnmower and the multifunctional lawnmower.*

H_{1a}. *There is a significant difference in the sound intensity level between a conventional lawnmower and the multifunctional lawnmower.*

The results of the two-sample *t*-test in Table 24 show that H_{0a} is rejected and H_{1a} is supported as the *p*-value is less than 0.05 [$t(39) = 127.62$, $p < 0.05$]. The mean sound intensity level for the multifunctional lawnmower is 58.87 dB which is significantly lower (about 34%) than the sound level produced by the conventional lawnmower (89.13 dB).

Table 24. Two-sample *t*-test for sound intensity level experiment.

Null hypothesis	H_{0a}: $\mu_1 - \mu_2 = 0$	
Alternative hypothesis	H_{1a}: $\mu_1 - \mu_2 \neq 0$	
t	df	p
127.62	39	0.000

This experiment is very similar to another experiment that observed the relation of noise from lawn maintenance machines with health risks [4]. Researchers who studied the noise levels around 3 m away from the machines found the ride-on mower to be the safest, with the lowest average noise level of 75 dB. In comparison, the multifunctional lawnmower in the present study emits a much lower sound intensity level, which validates its advantage in terms of sound comfort.

4.2. Cutting Ability Experiment

The objective of this experiment is to investigate whether there is a significant difference in grass height before and after the lawn is mowed by the multifunctional lawnmower. Thirty (30) square plots of grass (measuring 38 × 38 cm) are used in this experiment. Five (5) samples of grass height are measured and averaged for every plot.

A two-sample *t*-test is also used in this experiment. The null and alternative hypotheses for this test are stated below:

H_{0b}. *There is no significant difference in the grass height before and after the lawn is mowed by the multifunctional lawnmower.*

H_{1b}. *There is a significant difference in the grass height before and after the lawn is mowed by the multifunctional lawnmower.*

Table 25 shows that H_{0b} is rejected and H_{1b} is supported as the p-value is less than 0.05 [$t(40) = 11.11$, $p < 0.05$]. The mean grass height after the lawn was mowed is 3.36 cm which is significantly shorter (about 51%) than the grass height before the lawn was mowed (6.84 cm).

Table 25. Two-sample t-test for cutting ability experiment.

Null hypothesis	H_{0b}: $\mu_1 - \mu_2 = 0$	
Alternative hypothesis	H_{1b}: $\mu_1 - \mu_2 \neq 0$	
t	df	p
11.11	40	0.000

In association with the above findings, another study of lawn mowing and grass-cycling provided the authors with guidance on lawn maintenance particularly on how high to mow, when to mow and what to do with the grass clippings [32]. In a summary of proper mowing height for different types of grasses (see Table 26), it was found that the average proper mower height setting should be from 25.40 to 46.99 mm (2.54 to 4.70 cm). The mean cutting height for the present study's multifunctional lawnmower is around 3.36 cm which is within the suggested height setting range.

Table 26. Proper mowing height for different types of grasses [32].

Grass Type	Mower Height Setting (mm)	Mow When Grass at This Height (mm)
Bent grass	12.70–25.40	19.05–38.10
Bermuda grass (common)	25.40–38.10	38.10–57.15
Bermuda grass (hybrid)	12.70–25.40	19.05–38.10
Buffalo grass	25.40–50.80	38.10–76.20
Kentucky bluegrass	38.10–63.50	57.15–95.25
Kikuyu grass	25.40–38.10	38.10–57.15
Perennial grass	38.10–63.50	57.15–95.25
Tall fescue	38.10–76.20	57.15–114.3
St. Augustine grass	25.40–50.80	38.10–76.20
Zoysia grass	12.70–38.10	19.05–57.15
Average (mm)	25.40–46.99	38.10–70.49

It is important to note that this test verifies the ability of the prototype to perform its intended function, which is to shorten the grass height. The test does not necessarily prove that the prototype cuts grass better than conventional lawnmowers or existing smart lawnmowers. However, the literature support on grass height comparisons helps identify that the grass is shortened to an appropriate or acceptable height.

4.3. Polishing Performance Experiment

The objective of this experiment is to observe the performance of the multifunctional lawnmower's alternate function, which includes floor polishing. The experiment used mosaic tiles as the observed samples, and an illumination meter as the measurement instrument.

Using the illumination meter, a parameter known as illuminance (in lux) is measured. Table 27 shows the illuminance data for 3 different samples before and after the floor is polished. The illuminance of the floor increased by 20.7% after the polishing. Hence, the experiment concludes that there is a difference in illuminance before and after the floor is polished by the multifunctional lawnmower.

Table 27. Illuminance before and after the floor is polished.

Tile Sample	Illuminance (Lux)	
	Before	After
1	28.0	38.0
2	25.0	33.0
3	31.0	35.0
Mean	28.0	35.3

The present study's polishing experiment is similar to another research which investigated the effectiveness of banana peel as a substitute for commercial floor wax [33]. Instead of purely depending on observations, this prior research used a wide light range meter to measure the illuminance of different waxed floors for the comparison of shininess.

The results for shininess using the wide light range meter are shown in Table 28. Although not directly comparable in terms of results, this methodology is consistent with the one used in the present study since both performed a comparison on the shininess of floors through the measurement of illuminance.

Table 28. Shininess using wide light range meter [33].

Trial		Shininess (lux)	
		Using Banana Peel Floor Wax	Using Brand X Floor Wax
Ceramic Tile	1	16.6	16.3
	2	16.1	15.1
	3	16.1	15.2
Wood Parquet	1	12.4	11.1
	2	13.2	11.6
	3	12.1	11.3
Scarlet Oak	1	6.8	6.3
	2	7.4	6.6
	3	7.5	6.6
Mean		12.02	11.12

4.4. Battery Power Durability Experiment

A set of calculations are done to estimate the battery power durability of the multifunctional lawnmower. The operation time for two different scenarios (one with the solar charge controller and another without the solar charge controller) is calculated and shown in Table 29. Similar equations for charging and operation time have been used in another design project known as the Automated Mower Robo [26].

Table 29. Calculation of lawnmower operation time.

Operation Time (hours)	
Without Solar Charge Controller	Operation time (h) = Battery capacity (Ah)/Total current (A) Operation time (h) = 7.2/3.3 = 2.18 h
With Solar Charge Controller Assumption: effective charging time per day is 3 h	Extra battery capacity (Ah) = Effective charging time (h) × Solar panel total current Extra battery capacity (Ah) = 3 × (13.5/12) = 3.38 Ah Operation time (h) = (7.2 + 3.38)/3.3 = 3.21 h

Note: Ah stands for ampere-hours.

The total current was calculated as 3.3 A. With the battery capacity known, the operation time for the multifunctional lawnmower is calculated as 2.18 h (130.8 min). The peak sun hours in Malaysia occur from 12 to 3 p.m. (3 h) Thus, the solar panel would work best during this period. The solar panel used for the design is rated with 12 V, 4.5 W. Hence, the solar panel's total current is calculated by the total power in watts divided by the rated voltage in volts. Using the peak sun hours as the effective charging time, the extra battery capacity can then be calculated, followed by the calculation of the operation time. The operation time for the lawnmower with the solar charge controller is 3.21 h (192.6 min).

However, the recorded times for the operation with and without the solar charge controller are measured to be 118.5 min and 187.3 min respectively. The measured times are lesser than the calculated times by around 2 to 10%. The time difference between calculated and measured time may be due to the power consumption of other minor electronic parts such as the small bulbs on the microcontroller, motor controller modules, Bluetooth module and infrared sensors which are not considered in the calculation of operation time. The measured operation time with the solar charge controller is 36.73% longer than the operation time without the solar charge controller.

5. Conclusions

The aim of this study was to develop a novel, original, and multifunctional lawnmower through reviews of patent literature, research literature and variants of existing lawnmowers. The usability of the lawnmower was also successfully validated through data-driven experimentations.

The results from the sound intensity experiment showed that the average sound intensity level of the multifunctional robotic lawnmower is 58.87 dB, which is akin to that of a sound level in a library. This finding shows that the present study's lawnmower has the potential to reduce noise pollution and noise-induced health risks. In terms of multifunctionality, the prototype lawnmower also rendered the alternate function of a floor polisher and successfully improved the floor's shininess by about 20%.

The results of the battery power durability experiment showed that the operation time increased by around 37% when the solar panels were connected. The usage of solar panels not only improved the efficiency of the lawnmower, but also reduced non-renewable energy usage. This achievement is meritorious since Malaysia is still highly dependent on non-renewable resources for power supply with regard to the lawnmower industry.

Finally, the cost of producing a single prototype lawnmower is around 630 MYR. By considering the variable cost, fixed cost, break-even analysis and profit margin, the prototype can be priced at approximately 1000 MYR, which is about 10 times cheaper than other competitors in the market. On the whole, the multifunctional lawnmower proposed in this study is proven to be more effective than other conventional robotic lawnmowers with regard to its cost, multifunctionality and low reliance on non-renewable resources. A patent has been filled for this invention on 11 November 2020 with the patent application number PI2020005912. The invention's name is also known as the Multifunctional Robot.

For future research, more tests can be done on the lawnmower's locomotion ability on different terrains and cutting ability for different types of grass. In addition, advanced problem-solving tools such as TRIZ or Design Thinking can be used to further develop the multifunctional lawnmower in terms of its degree of inventiveness. Lastly, further analysis pertaining to the invention's polishing attribute (i.e., the polisher's design requirements, design conceptualisation and selection, stress analysis and material selection) can be done for future research.

Although more studies are required to fully develop a commercial-ready multifunctional lawnmower, this study not only opens an avenue for researchers to conduct more investigations on smart lawn mowing machines, but also assists Malaysia in her journey towards embracing the fourth industrial revolution.

Author Contributions: Conceptualization, C.Q.K. and P.K.N.; Data curation, C.Q.K.; Formal analysis, C.Q.K. and K.W.L.; Investigation, C.Q.K. and P.K.N.; Methodology, C.Q.K., P.K.N. and K.W.L.; Project administration, P.K.N.; Resources, P.K.N.; Supervision, P.K.N. and K.W.L.; Validation, C.Q.K.; Writing—original draft, C.Q.K. and P.K.N.; Writing—review & editing, C.Q.K., P.K.N. and K.W.L. All authors have read and agreed to the published version of the manuscript.

Funding: This research received no external funding.

Institutional Review Board Statement: Not applicable.

Informed Consent Statement: Not applicable.

Data Availability Statement: This project contains the following underlying data: Data Availability Sheet.docx (stress analysis for preliminary design, experimental data, program codes and circuit diagram). The data can be found at Figshare: doi.org/10.6084/m9.figshare.14617122 (accessed on 21 May 2021). Data are available under the terms of the Creative Commons Attribution 4.0 International license (CC-BY 4.0).

Acknowledgments: The researchers gratefully thank the Faculty of Engineering and Technology as well as Multimedia University for their support in allowing this research to be carried out. The researchers also thank Chi Keng Kang for his useful comments on matters related to this research project.

Conflicts of Interest: The authors report no conflict of interest.

References

1. Khodke, K.R.; Kukreja, H.; Kotekar, S. Literature Review of Grass Cutter Machine. *Int. J. Emerg. Technol. Eng. Res.* **2018**, *6*, 97–101.
2. Calculating Mowing Times and Productivity. Encore Power Equipment. 2012. Available online: http://www.encoreequipment.com/wordpress/wp-content/uploads/2012/06/Mowing-Times-and-Productivity.pdf (accessed on 17 November 2019).
3. Banks, J.L. National Emissions from Lawn and Garden Equipment. 2015. Available online: http://www.arb.ca.gov/msei/offroad/pubs/offroad_overview.pdf (accessed on 17 November 2019).
4. Tint, P.; Tarmas, G.; Koppel, T.; Renihold, K.; Kalle, S. Vibration and Voice Caused by Lawn Maintenance Machines in Association with Risk to Health. *Agron. Res. Biosyst. Eng.* **2012**, *10*, 251–260.
5. Niquette, P.A. Noise exposure: Explanation of OSHA and NIOSH safe exposure limits and the importance of noise dosimetry. *Can. Hear. Rep.* **2012**, *9*, 22–29.
6. Harris, C.; Madonick, J.; Hartka, T.R. Lawn mower injuries presenting to the emergency department: 2005 to 2015. *Am. J. Emerg. Med.* **2018**, *36*, 1565–1569. [CrossRef] [PubMed]
7. Ren, K.S.; Chounthirath, T.; Yang, J.; Friedenberg, L.; Smith, G.A. Children treated for lawn mower-related injuries in US emergency departments, 1990–2014. *Am. J. Emerg. Med.* **2017**, *35*, 893–898. [CrossRef] [PubMed]
8. Kadam, S.; Bhandalkar, A. Robotic Lawn Mower Market by Range, End User, and Distribution Channel: Global Opportunity Analysis and Industry Forecast, 2018–2025. 2018. Available online: https://www.alliedmarketresearch.com/press-release/robotic-lawn-mower-market.html (accessed on 19 November 2019).
9. Samsudin, M.S.N.; Rahman, M.M.; Wahid, M.A. Power Generation Sources in Malaysia: Status and Prospects for Sustainable Development. *J. Adv. Rev. Sci. Res.* **2016**, *25*, 11–28.
10. Lim, S.; Ng, P. Synthesisation of Design Features for Multifunctional Stretcher Concepts. *J. Med. Eng. Technol.* **2020**, *45*, 145–157. [CrossRef] [PubMed]
11. Bruzzone, L.; Quaglia, G. Review article: Locomotion systems for ground mobile robots in unstructured environments. *Mech. Sci.* **2012**, *3*, 49–62. [CrossRef]
12. Morin, P.; Samson, C. *Motion Control of Wheeled Mobile Robots*; Springer: Berlin, Germany, 2008.
13. Parmar, J.J.; Savant, C.V. Selection of Wheels in Robotics. *Int. J. Sci. Eng. Res.* **2014**, *5*, 339–343.
14. Martínez, J.L.; Mandow, A.; Morales, J.; Pedraza, S.; García-Cerezo, A. Approximating kinematics for tracked mobile robots. *Int. J. Rob. Res.* **2005**, *24*, 867–878. [CrossRef]
15. Zamanov, V.; Dimitrov, A. *Tracked Locomotion and Manipulation Robots*; Bulgarian Academy of Sciences: Sofia, Bulgaria, 2012.
16. Lokhande, N.G.; Emche, V.B. Mechanical Spider by Using Klann Mechanism. *Int. J. Mech. Eng. Comput. Appl.* **2013**, *1*, 13–16.
17. Brooks, R.A.; Flynn, A.M. Fast, cheap, and out of control. *J. Br. Interplanet. Soc.* **1989**, *42*, 478–485. [CrossRef]
18. Pavlak, A. *Material Selection Analysis for the Development of an Integrated Surface Vehicle System*; Southern Illinois University: Carbondale, IL, USA, 2016.
19. Ashby, M.F. *Materials Selection in Mechanical Design*, 3rd ed.; Jordan, H., Ed.; Butterworth Heinemann-Elsevier Science: Oxford, UK, 2005.
20. Güven, Y.; Coşgun, E.; Kocaoğlu, S.; Gezici, H.; Yilmazlar, E. Understanding the Concept of Microcontroller Based Systems to Choose the Best Hardware for Applications. *Res. Inven. Int. J. Eng. Sci.* **2017**, *7*, 38.
21. Rajan, C.; Megala, B.; Nandhini, A.; Priya, C.R. A Review: Comparative Analysis of Arduino Micro Controllers in Robotic Car. *Int. J. Mech. Aerosp. Ind. Mechatron. Eng.* **2015**, *9*, 371–380.
22. Pujari, S.; Panda, A.; Muduli, P.P.; Badhai, R.; Nayak, S.; Sahoo, Y. A Learning Model for 8051 Microcontroller Case Study on Closed Loop DC Motor Speed Control. *Int. J. Emerg. Technol. Adv. Eng.* **2013**, *3*, 2250–2459.
23. Willis, H.J. Automated Lawn Cutting and Vacuum System. U.S. Patent US7185478B1, 6 March 2007.
24. Martin, R.L. Automated Lawn Mower or Floor Polisher. U.S. Patent US4887415A, 19 December 1989.
25. Ruffner, B.J. Multifunctional Mobile Appliance. U.S. Patent US20020049522A1, 31 December 2002.

26. Tanaji, S.V.; Chandrakant, C.S.; Shashikant, P.S.; Raju, G.O.; Bhalchandra, G.S. Automated mower robo. *Int. Res. J. Eng. Technol.* **2018**, *5*, 1–4.
27. Dewangan, A.K.; Raja, R.; Singh, R. Multi Sensor and Multifunctional Robot With Image Mosaic. *Int. J. Sci. Eng. Technol. Res.* **2014**, *3*, 677–680.
28. Wójcik, K. The Influence of the Cutting Attachment on Vibrations Emitted by Brush Cutters and Grass Trimmers. *For. Res. Pap.* **2015**, *76*, 331–340. [CrossRef]
29. Halley, M. Robomow RX12u Review. 2018. Available online: https://easylawnmowing.co.uk/robomow-rx12u-review/ (accessed on 19 November 2019).
30. Halley, M. Husqvarna 105 Automower Review. 2016. Available online: https://easylawnmowing.co.uk/husqvarna-105-automower-review/ (accessed on 19 November 2019).
31. Hales, D. iRobot Terra t7 Review. 2019. Available online: https://moderncastle.com/irobot-terra-t7-review/ (accessed on 19 November 2019).
32. Harivandi, A.; Gibeault, V.A. *Mowing Your Lawn and Grasscycling*, 1st ed.; Division of Agriculture and Natural Resources, Regents of the University of California: Oakland, CA, USA, 1999.
33. Feliciano, P.A.C.; Ong, K.A.G. Effectiveness of Musa Paradisiaca (Banana) Peel as an Alternative to Commercial Floor Wax for Household Use in the Philippines. *Asia Pac. J. Multidiscip. Res.* **2019**, *7*, 38–48.

Article

HIGROTERM: An Open-Source and Low-Cost Temperature and Humidity Monitoring System for Laboratory Applications

Renan Rocha Ribeiro *, Elton Bauer and Rodrigo Lameiras 

Department of Civil and Environmental Engineering, University of Brasília, Brasilia 70910-900, Brazil; elbauerlem@gmail.com (E.B.); rmlameiras@gmail.com (R.L.)
* Correspondence: renan.rocha.ribeir@gmail.com

Abstract: Low-cost electronics developed on easy-to-use prototyping platforms, such as Arduino, are becoming increasingly popular in various fields of science. This article presents an open-source and low-cost eight-channel data-logging system for temperature and humidity monitoring based on DHT22 (AM2302) sensors, named HIGROTERM. The system was designed to solve real needs of the Laboratory of Material Testing of the Department of Civil and Environmental Engineering at the University of Brasília. The system design, functionalities, hardware components, source code, bill of materials, assemblage and enclosure are thoroughly described to enable complete reproduction by the interested reader. The terminologies and instructions presented were simplified as much as possible to make it accessible to the greatest extent to researchers from different areas, especially those without electronics background. The data-acquisition system has an estimated total cost of USD 96.00, or USD 136.00 if eight sensor nodes are included, with a considerable margin for cost reduction. The authors expect that the HIGROTERM system may both be a valuable low-cost and customizable tool for the readers, as well a source of innovation and interest in low-cost electronics for real problem-solving in various fields of science.

Keywords: temperature; humidity; low-cost; open-source; Arduino; data-logging; material testing; laboratory; electronics

Citation: Rocha Ribeiro, R.; Bauer, E.; Lameiras, R. HIGROTERM: An Open-Source and Low-Cost Temperature and Humidity Monitoring System for Laboratory Applications. *Inventions* **2021**, *6*, 84. https://doi.org/10.3390/inventions6040084

Academic Editors: Francisco Manzano-Agugliaro, Luigi Fortuna and Tek-Tjing Lie

Received: 28 July 2021
Accepted: 21 September 2021
Published: 15 November 2021

1. Introduction

The use of low-cost, open-source electronics prototyping, and do-it-yourself (DIY) platforms, has become increasingly common across all fields of science, in which needs for tailor-made solutions in budget-tight situations are recurrent. The Arduino platform [1] is among the most popular electronics prototyping platforms, with myriad applications reported in the literature, such as smartphone-integrated systems for environmental measurement [2], agricultural applications [3–5], landslide monitoring [6], meteorological applications [7], systems for indoor environmental monitoring and control with various types of sensors [8–11], Internet-of-Things (IoT) temperature and humidity systems [12], temperature and humidity measurements inside concrete structures [13], air quality monitoring [14], and custom-made printed circuit boards (PCBs) for various laboratorial automation [15].

This work introduces an open-source, low-cost, eight-channel data-logging system for temperature and humidity monitoring named HIGROTERM, based on DHT22 (AM2302) sensors [16], describing its hardware components, assembly process, source code development, operation guidelines and an example of application.

The HIGROTERM system was developed in the context of real needs found in the Laboratory of Material Testing of the Department of Civil and Environmental Engineering at the University of Brasília, for teaching and researching activities. The need to monitor the temperature and humidity, two of the main environmental variables in building material studies, at many points in several environments simultaneously was common in many research projects conducted at the laboratory, being requirements of many standardized tests [17–20]. The environments also varied in size and location: it could be the monitorization of several points across an entire room or many small experimental chambers or boxes located in a workbench. In this way, a multichannel sensor data logger system, capable of measuring temperature and humidity from small sensor nodes, with flexible deployment, seemed to be the appropriate solution to these problems.

Commercial off-the-shelf solutions for temperature and humidity monitoring were considered inefficient, due to either high costs of multichannel data logger systems and sensors, or the impracticability of usual single stand-alone remote sensors, which may be too bulky to monitor small environments, or rely on batteries that run out too quickly. An attempt was made to write this paper as accessible as possible to non-experts in electronics, so as to contribute not only to the widespread use of HIGROTERM systems, but also of prototyping electronics systems for real problem solving in different scientific communities.

2. Materials and Methods

2.1. System Requirements and Design

The HIGROTERM system was designed to be a multichannel data logger for temperature and humidity sensors that meets some basic requirements, devised to aid the task of monitoring and controlling laboratorial environments. The following requirements were outlined for the system:

1. The system should be, at the greatest extent possible, developed in open source and easy-to-use platform to favor the widespread across the scientific community;
2. The system's hardware and software were to be both low-cost because budget limitations are common in research projects;
3. At least eight simultaneous sensors' channels were to be supported by the system, because it was deemed sufficient for the current needs at the Laboratory of Material Testing at the University of Brasília, but its design should enable the easy addition of more channels;
4. A tailor-made user interface was to be incorporated to the system, based on a liquid-crystal display (LCD) touch screen that allows the user to easily configure the system in an on-demand approach, without the need for any additional equipment, such as a computer, as commonly found with other systems reported in the literature [8–10,12,15];
5. The collected data were to be stored in physical media with wide compatibility with other systems, to facilitate data transfer;
6. The date and time of each sample was also to be recorded, to enable reconstruction of the data history;
7. The system would be powered by a wall socket because such availability is reasonable in a laboratorial environment and ensures a long-term running system.

In order to quickly obtain a first working prototype, the HIGROTERM system was built using the Arduino platform, a low-cost and easy-to-use prototyping hardware and software platform for electronic systems [1]. The Arduino platform comprises several types of prototyping boards, each containing one microcontroller unit (MCU) and the circuitry necessary for peripheral controls and interfaces, and to upload source codes by using a computer. The boards are conceived in a way that allows the building of electronic control systems to be virtually reduced to a simply plug-and-play hardware process. The platform also provides an integrated development environment (IDE) for programming the MCUs contained in all Arduino boards with a C++/C based language. Many ready-to-use source codes are available online, and there are many solid online support communities. In this way, the development of electronic systems using Arduino boards is simplified and accessible for users lacking a strong electronics and programming background.

Among the available prototyping boards in the Arduino platform, the Arduino MEGA board was selected for the development of the system [21]. The MEGA board can be considered an intermediate-performance board in the Arduino platform, with an ATmega2560 MCU of 16 MHz [22], in a board with 16 analog pins, 54 digital pins, 4 kB of EEPROM memory, 8 kB of RAM memory, and 256 kB of Flash memory. The MEGA board is internally operated in 5 V and can be powered with a 7–12 V external power source. The choice of the MEGA board was driven by the system's requirements: (i) #3, which required large number of digital pins for multiple temperature and humidity sensor channels; and (ii) #4, because the MEGA board has off-the-shelf interface solutions for LCD touch screens based on shields, which are boards designed to be directly plugged on the top of Arduino boards, in a plug-and-play fashion, without requiring use of soldering or other type of specialized connections [23] which, thus, also comply with the easy-to-use system's requirements #1.

For complying with the system requirements #5, #6 and #7, SD and Real-Time Clock modules compatible with the Arduino platform were used. These modules, similarly, to the LCD touch screen shield, work in a plug-and-play fashion.

2.2. Overall System Design

In its first version, the system was mounted with jumper cables in a solderless breadboard, for easier prototyping. Figure 1 shows the HIGROTERM system prototype, in its current development state. In the figure, the jumper cables are not fully shown for better clarity, and the same colors are used for cables that are connected to the same trail lines.

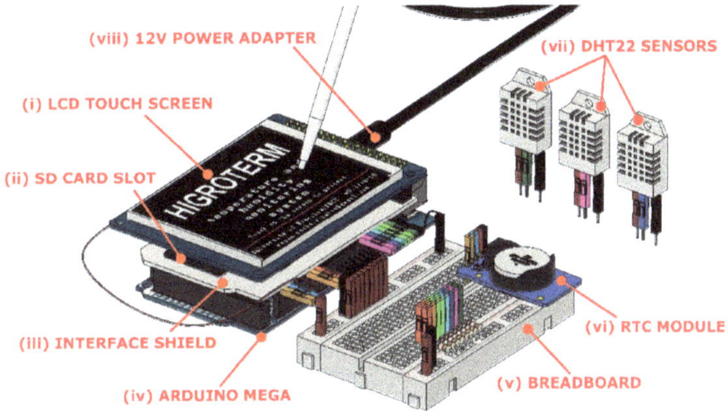

Figure 1. HIGROTERM system with its main parts.

The main parts of the system, as indicated in the figure, are: (i) an LCD touch screen for user interfacing with the system; (ii) an SD card slot, which is, in fact, integrated on the LCD touch screen shield; (iii) an intermediate interface shield, which allows connecting the LCD touch screen shield, operated internally at 3.3 V, directly to the Arduino MEGA board, operated internally at 5 V; (iv) the Arduino MEGA board; (v) the solderless breadboard; (vi) a Real-Time Clock (RTC) module, which is able to precisely track the time and date, even if a power shortage occurs; (vii) DHT22 sensors for temperature and humidity measurements; and (viii) a 12 V power wall socket. Further details about the hardware components are presented in the following sections.

Once the system was fully functional, the solderless breadboard was substituted with a solderable breadboard, in which the jumper cables and other components were directly soldered. This was performed to increase the system reliability under use, because the solderless breadboard would be prone to short circuitry or the disconnection of cables if the system was mishandled.

3. Results and Discussions

The HIGROTERM source code, hardware design, schematics, and other useful documents for mounting and usage, are available in its GitHub repository [24]. All the related documents for the HIGROTERM system have been released under the terms of the MIT license [25], chosen so as to provide the system with the fewest restrictions possible. However, the HIGROTERM source code uses external libraries developed by third parties, which are released under myriad different licenses. The most restrictive among them is the CC BY-NC-SA 4.0, which does not allow commercial use [26]; therefore, readers interested in using the HIGROTERM system should comply with the implications of this licensing.

3.1. Hardware and Electronic Schematic

Figure 2 presents a general schematic, in an exploded view, of the hardware components of the HIGROTERM system and how they are assembled. In this figure, the solderable breadboard, used in the final version of the system, is presented, instead of the solderless breadboard illustrated in Figure 1 with the use of the solderable perfboard and how they assemble. The blue arrows indicate how solderless pieces fit together. The components are numerically identified to provide better reference in the text. Additionally, for the sake of clarity, just one DHT22 sensor is shown in Figure 2, although the (viii) sensor connections for the other seven sensors are still illustrated in terms of connectors soldered on the solderable perfboard.

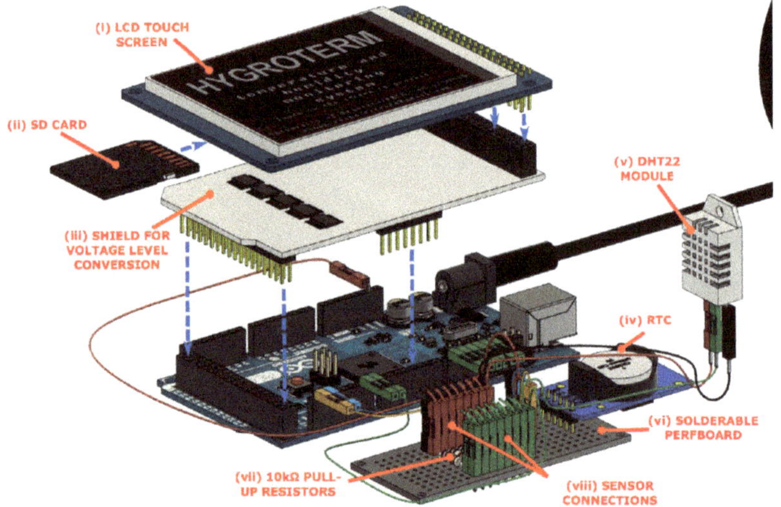

Figure 2. HIGROTERM system with its main parts and how they assemble.

The (i) LCD touch screen shield used was from the TFT_320QDT_9341 model, which comprised a single printed circuit board (PCB) containing: a 3.2-inch, 240 by 320 pixels thin-film transistor (TFT) LCD screen controlled by an integrated ILI9341 controller; a resistive touch screen controlled by an integrated XPT2046 controller; and a (ii) standard SD card socket. This PCB operated at 3.3 V and contained 40 pins that were used to interface with the MCU. The Arduino MEGA works internally at 5 V; therefore, a voltage level conversion was required on each of the 40 pins of the TFT_320QDT_9341 PCB. This task was performed by the intermediate (iii) voltage level conversion shield, from the TFT LCD Mega Shield V2.2 model. This shield transferred the 40 pins of the LCD PCB to 50 pins distributed along the Arduino MEGA board layout. As shown in Figure 2, these components were simply stacked together on top of the Arduino MEGA board, and no soldering was required to produce reliable connections.

The (iv) RTC module used is based on a DS3231 chip [27], built in a ready-to-use module, and the temperature and humidity sensor was a capacitive-type (v) DHT22 module, which is also identified as AM2302 [16]. To provide a reliable connection between these components and the Arduino MEGA board, a (vi) solderable perfboard was used as an intermediate connection medium. In it, jumper cables sockets were soldered to receive jumper cables that came from the MEGA board and from the components. In addition to improving the circuit organization and reliability, the solderable perfboard aided the connection of (vii) 10 kΩ pull-up resistors, i.e., resistors between the power and data lines, required by the DHT22 modules.

The DHT22 connections to the Arduino MEGA board were performed by three-pin screw male-female 2EDG model terminal connectors, for better system flexibility and reliability. In this way, the Arduino MEGA pins were connected to 2EDG female connectors, and three wire cables, produced with 2EDG male connectors at one end and female Molex-type KK 254 connectors at the other end, were used to connect the system to the DHT22 sensors, which were soldered to male Molex-type KK 254 connectors. Hot glue was applied to all exposed contacts to avoid short circuits and disconnections. This solution provided flexibility for the system, allowing the use of different cables, of different lengths, to connect the sensors for various purposes and monitoring tasks, while also enabling the future replacement of damaged sensors without requiring any modification to the system and cables themselves.

To facilitate the comprehension of the system wiring, Figures 3 and 4 present an overview of the system wiring in a solderless breadboard and the electronic schematic, respectively. The LCD touch screen is the only component not shown, because it simply stacks upon TFT LCD Mega Shield V2.2 model, not requiring further details about its connection. The schematic is based on the pin numbering used on the source code available on its GitHub repository [24].

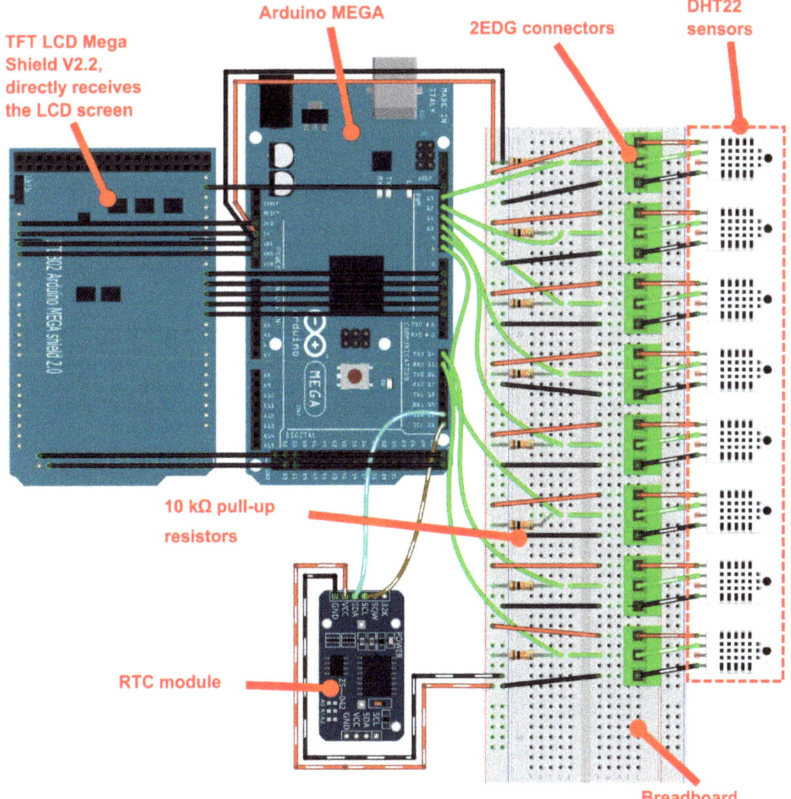

Figure 3. Overview of the system wiring using a solderless breadboard.

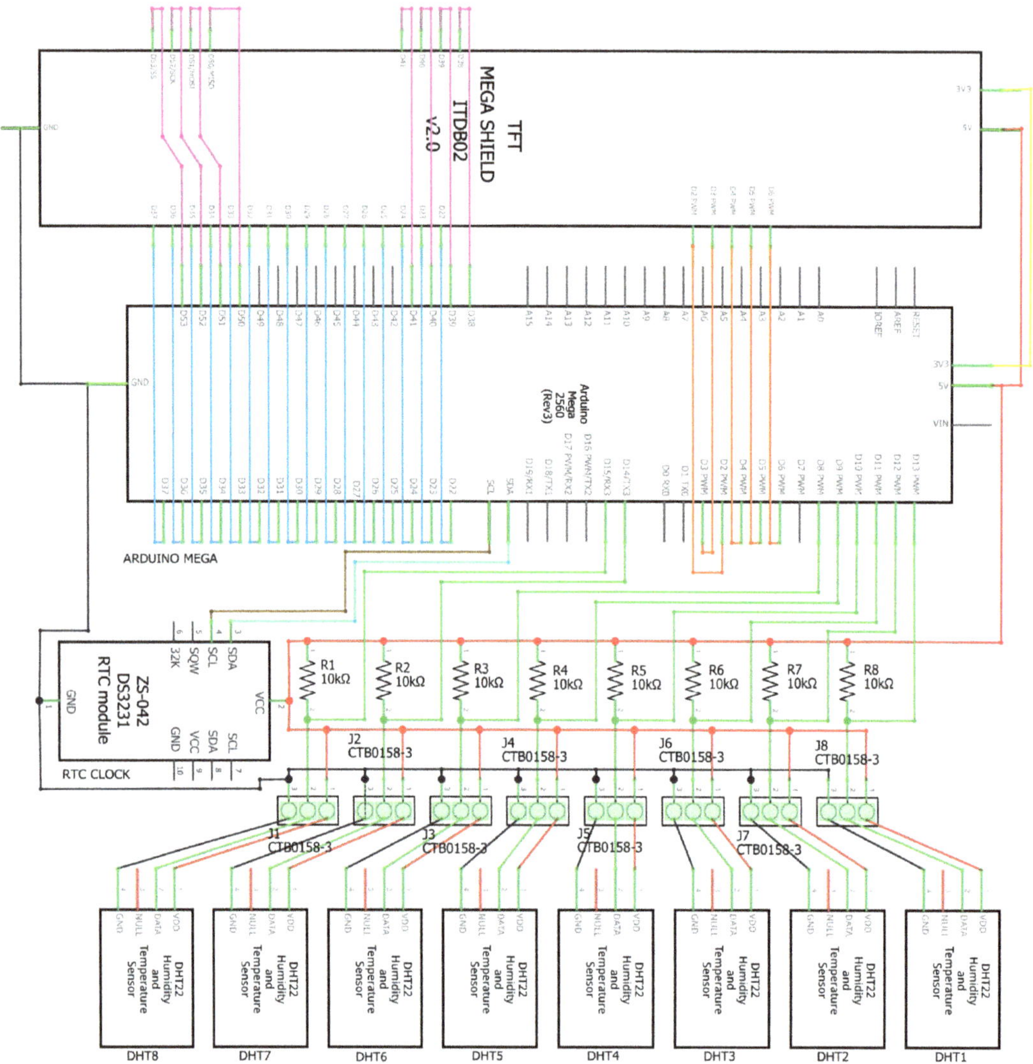

Figure 4. Electronic schematic.

The RTC module required four connection wires: a 5 V power and ground wires, for powering up the module; and SDA and SCL wires connected to the SDA and SCL pins of the Arduino MEGA, i.e., digital pins 20 and 21, respectively, for data transmission using the I2C protocol.

The DHT22 modules required three connection wires: a 5 V power and ground wires, for powering up the module; and a wire connected to an MCU input/output pin for data transfer, because this module uses a custom single-wire protocol. The protocol was based on digital signals; therefore, the module's data pin should preferentially be connected to an Arduino's digital pin, although analog pins can also be used.

For more practicality in mounting the system, a printed circuit board (PCB) was designed, as illustrated in Figure 5. This PCB required only direct soldering of the Arduino MEGA board to it, which may be performed with stackable pin headers instead of the traditional headers, the 2EDG female connectors and the RTC clock module. The use of this PCB required no jumper cables and allowed the system to be compacted into a 150 mm long, 80 mm wide, and approximately 50 mm tall box. The schematic of this PCB in Gerber format, which is ready to be sent for the PCB production, may be downloaded from the HIGROTERM official GitHub repository [24].

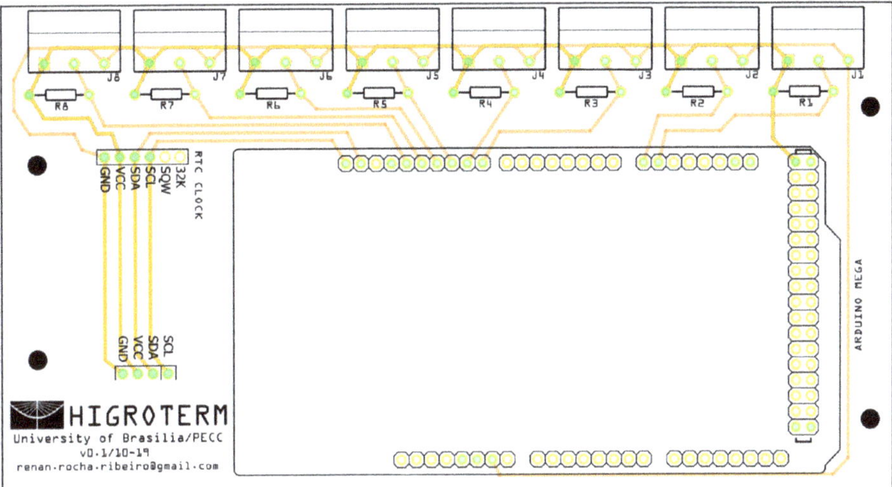

Figure 5. Electronic schematic.

3.2. Source Code

The Arduino platform allows a microcontroller unit (MCU) to be programmed using the Arduino IDE (integrated development environment), which uses a C++/C-based language and is currently available for Windows, Linux, Mac OS and as an online Web Editor [28]. Using this IDE, the user can write, compile and upload the source code to the MCU if the Arduino board is connected to the computer via a USB port. The source code implemented on the IDE is saved under the *ino* format, which is called an Arduino File.

This format allows the source code to be divided, for better organization, in multiple Arduino Files, which are sequentially appended during compilation accordingly to the alphabetical order of their file names. This strategy was adopted in the HIGROTERM source code, dividing its 3080 lines of code into 37 Arduino Files. These files were named with an alphanumeric prefix so as to maintain the correct order for compilation. Additionally, alphanumeric naming was used to create file groups that refer to correlated functions, aiding navigation throughout the code and future implementations.

Figure 6 presents the flowchart of the source code algorithm. The source code was designed based on the navigation of the user through a set of six screens, indicated as dashed-line background boxes in the flowchart. Each screen is defined by its graphical elements and a set of actions the user can undertake while in it. Figure 7 presents the six screens, as well as the three possible states of the Monitoring Table Screen, in which the data logging and visualization occurs, which thus constitutes the main screen of the HIGROTERM system.

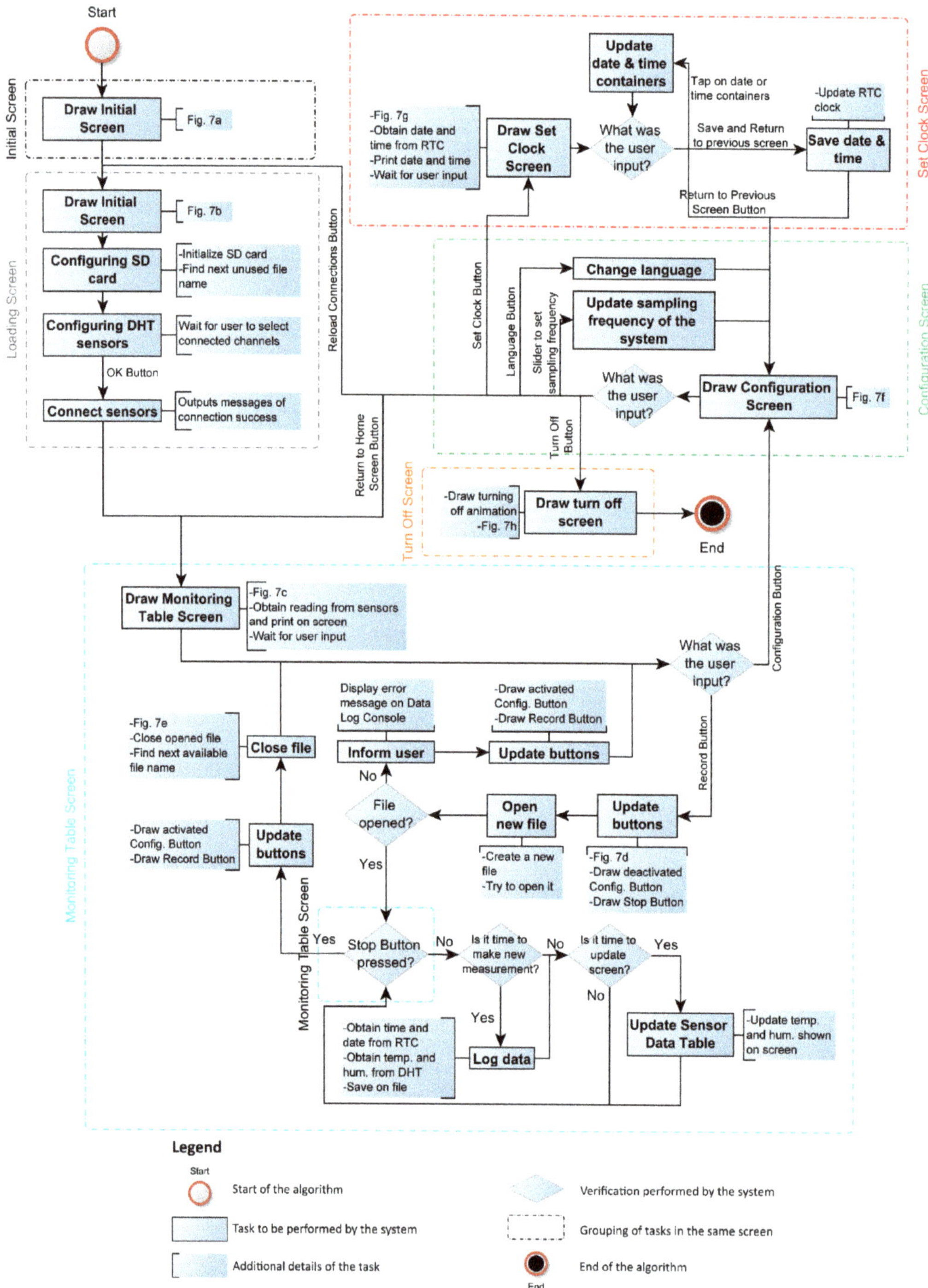

Figure 6. Source code algorithm flowchart.

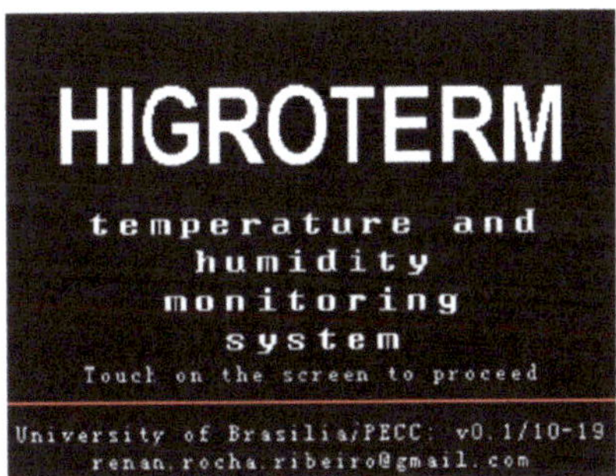

(**a**) Initial Screen

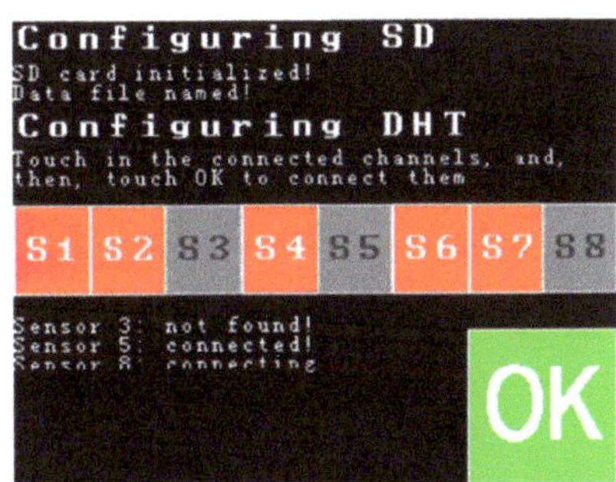

(**b**) Loading Screen

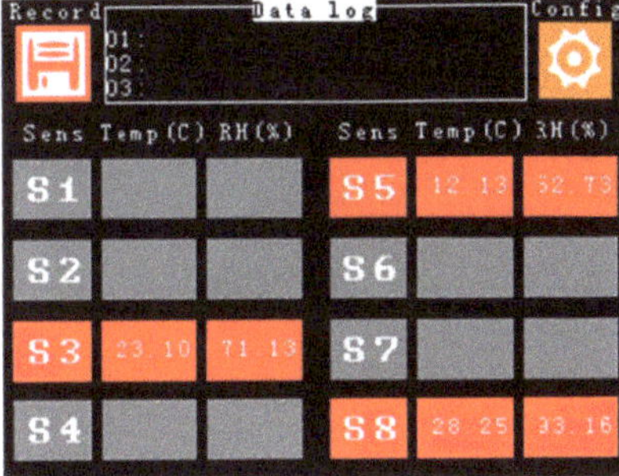

(**c**) Monitoring Table Screen. State 1: entering this screen from the Loading Screen.

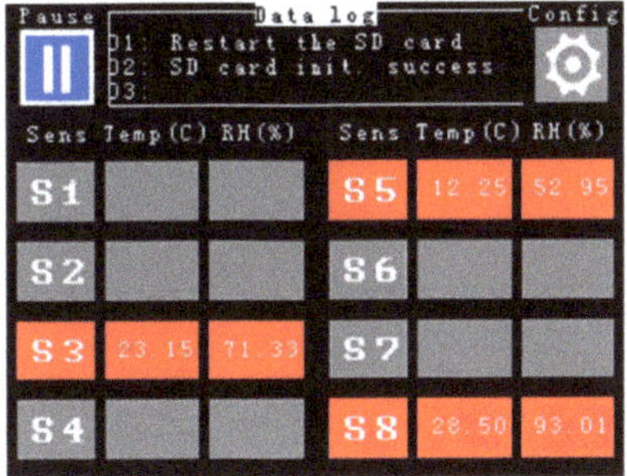

(**d**) Monitoring Table Screen. State 2: during a sampling cycle.

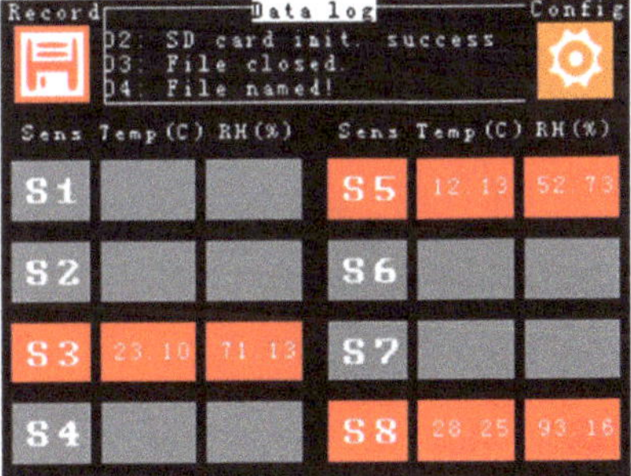

(**e**) Monitoring Table Screen. State 3: after a sampling cycle was stopped.

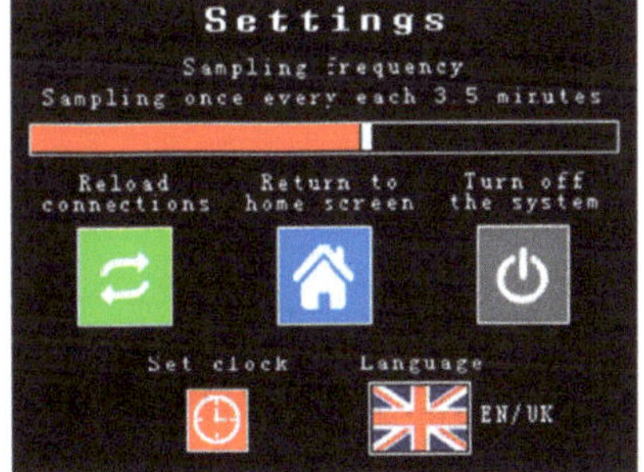

(**f**) Configuration Screen.

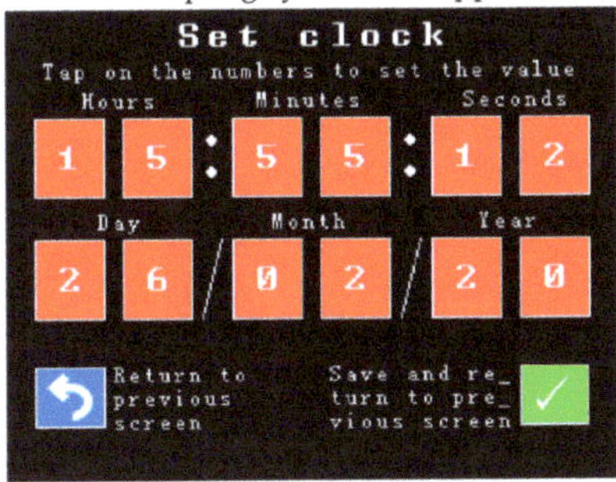

(**g**) Set Clock Screen.

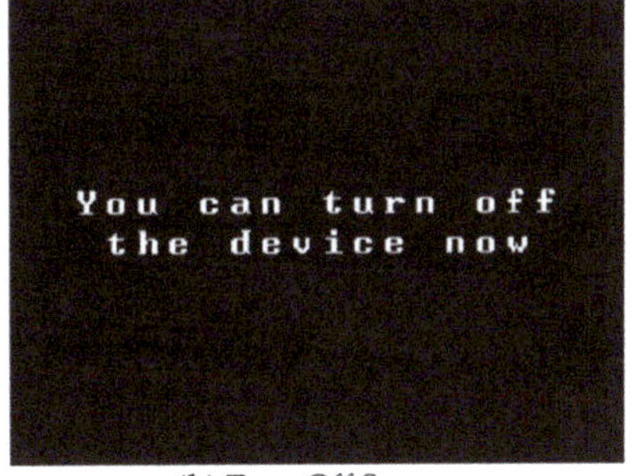

(**h**) Turn Off Screen.

Figure 7. HIGROTERM system screens.

3.3. Enclosure Case

The whole system was enclosed in a rigid plastic case, as shown in Figure 8. The option for a case led to greater user-friendliness, because systems built in a raw prototype fashion, with exposed cables and contacts, may be intimidating for users with no background in electronics, while also being more prone to damages to the circuitry. The case comprised an off-the-shelf rigid plastic case for general-purpose electronics, with perforations made with a high-speed rotary drill to enable mounting of the system. Although being a very accessible option, in economic and technical terms, for creating the system enclosure, this technique resulted in less-than-perfect finishing due to the difficulty in performing these tailor-made perforations.

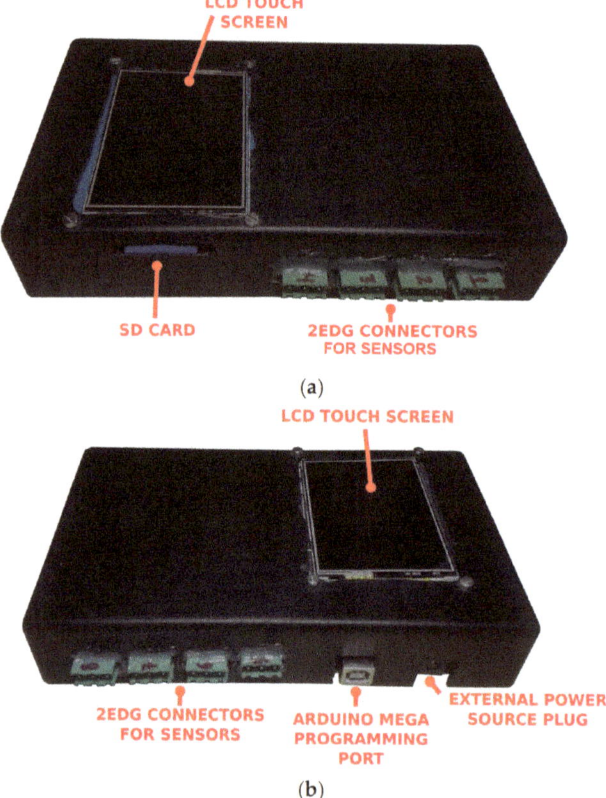

Figure 8. HIGROTERM enclosure in a plastic box: (**a**) front view with SD card and sensor; (**b**) back view with sensor connectors, programming port and power source plug.

To increase the quality of the enclosure finishing, 3D printing, or plastic injection manufacturing techniques may be used either for fabricating the whole enclosure or just for finish frames to be inserted around the perforations made. The LCD touch screen and Arduino MEGA board were screwed to the plastic case, while the 2EDG connectors were fixed with hot glue on the openings made. Holes were made to enable connection of an external power source plug and access of the Arduino Mega programming port, via USB, which enabled easy update of the system source code.

The inside of the HIGROTERM case, showing the hardware component connections, is shown in Figure 9. In Figure 2, for better image clarity, the RTC module is shown directly attached to the solderable perfboard. In fact, the RTC module was connected to the perfboard with jumper wires as well, similarly to the 2EDG and Arduino MEGA board, so as to better accommodate the system inside the available space of the plastic enclosure box, as shown in Figure 9.

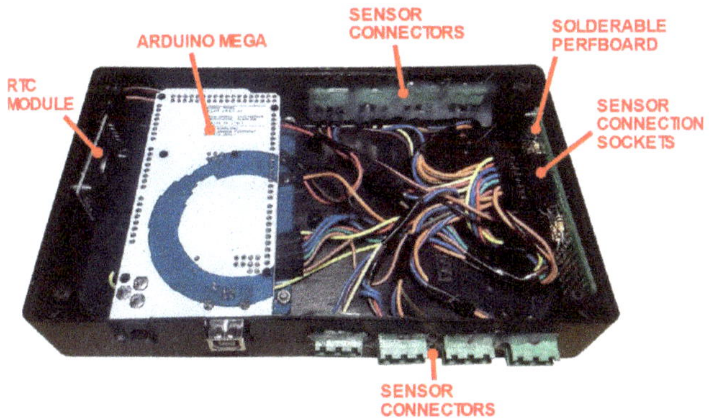

Figure 9. HIGROTERM hardware components.

3.4. Bill of Materials

Table 1 presents the bill of materials required to build one unit of the HIGROTERM system, with eight sensor channels, including the sensors themselves. It was estimated to need a 1.25 m of cable per sensor, which resulted in a total of 10 m for all eight sensors. The estimated costs were taken randomly from different websites and may vary significantly, and shall be used only as guidance on the overall cost, because time and place factors may greatly influence the values listed.

Table 1. Bill of materials and estimations of costs.

Component	Quantity	Unit	Price (USD)
Arduino MEGA board	1	Un.	15.00
3.2″ 240 × 320 LCD touch screen model TFT_320QDT_9341	1	Un.	12.00
Voltage-level conversion shirl model—TFT LCD Mega Shield V2.2	1	Un.	2.00
8 GB microSD card	1	Un.	11.00
12 V power supply adapter	1	Un.	6.00
RTC DS3231 module	1	Un.	3.00
DHT module	8	Un.	40.00
Solderable perfboard	1	Un.	4.00
10 kΩ resistor	8	Un.	8.00
2.54 mm jumper cables (male-male)—pack with 40 units	1	Pack	1.00
2.54 mm jumper cables (male-female)—pack with 40 units	1	Pack	1.00
2.54 mm female pin socket connector	40	Un.	6.00
2EDG connector	8	Un.	4.00
4-wire AWG cable to connect sensors to system	10	Meter	14.0
Rectangular rigid plastic enclosure box	1	Un.	10.00
Total (USD)		136.00	

The total cost of USD 136.00, with all sensors included, demonstrates that the low-cost requirement was achieved, because this value is many times lower than currently available commercial solutions for eight-channel temperature and humidity monitoring systems.

Considering that almost 30% of this cost is only the sensor modules, the data-acquisition system can certainly be considered as low-cost, with an estimated cost of USD 96.00. Additionally, as the number of channels increased, the low-cost advantage is even more evident, because additional channels can be added with little effort and small increases in costs.

3.5. Validation

In order to preliminarily validate the performance of the HIGROTERM system, simultaneous measurements on the same environment were performed with the system, and a commercial digital thermohygrometer, model K29-5070H manufactured by Kasvi, whose main characteristics, accordingly to the manufacturer, are: internal and external measurements via measurement probe; temperature range of −10 °C to +50 °C for internal measurements, and −50 °C to +70 °C for external measurements; humidity range of 20% to 99% RH, and accuracy of 1 °C and 5% RH [29]. Comparatively, the DHT22 sensors used with the HIGROTERM system have the following characteristics, accordingly to the manufacturer: temperature range of −40 °C to +80 °C; humidity range of 0% to 100% RH, accuracy of 0.5 °C and 5% RH, and precision ±1%RH and ±0.2 °C [16].

Commercial digital thermohygrometers are the go-to solution adopted so far in research and teaching activities in the Laboratory of Material Testing; therefore, its measurements were considered as references. Both sensors were placed in the same environment, which had average temperature and humidity during the test, of 26.4 °C and 28%, respectively, and 10 measurement sessions were performed, approximately every 5 min with each sensor. The measurements of the commercial system were composed of a single sample manually registered from the display value, because the system does not have a logging capability, whereas the measurements of the HIGROTERM system were composed of 10 samples taken every 2 s, so both accuracy and precision could be evaluated. The temperature and humidity errors were computed as:

$$error = \left| \overline{x}_{higroterm} - x_{commercial} \right| \quad (1)$$

In which || denotes the modulus or absolute value of the argument, $\overline{x}_{higroterm}$ is the average temperature or humidity obtained from the 10 samples taken by the HIGROTERM system, and is the temperature or humidity measured by the commercial reference system. Additionally, the standard deviation of the temperature and humidity measurements of the HIGROTERM system were computed, as an estimate of the system's precision.

Figure 10 presents the results of the validation in terms of error and standard deviation observed in each of the 10 measurement sessions of temperature and humidity.

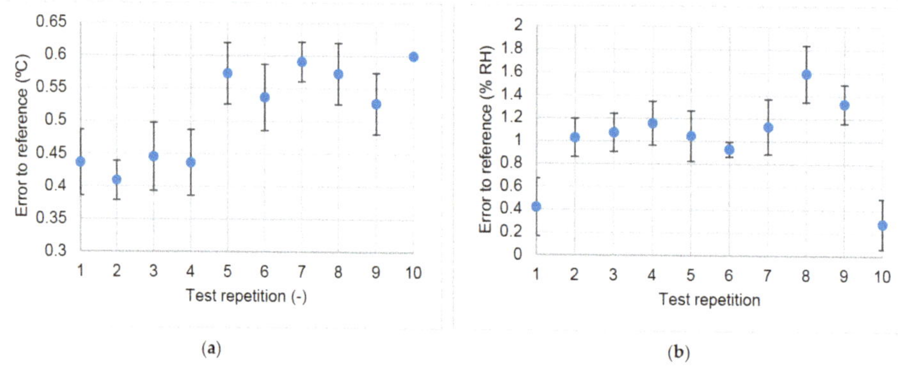

Figure 10. Validation results: (**a**) temperature measurements; (**b**) relative humidity measurements. The results are given in terms of error of the average measurement value to the reference values and variation of the samples in terms of one standard deviation above and below the average.

The validation results indicated that the accuracies obtained were very close to those informed by the DHT22 manufacturer, with values observed within 0.4–0.6 °C against an accuracy of 0.5 °C, for temperature, and 0.3%–1.6%, against 5%, for relative humidity, considering the commercial device as a reference. Moreover, the standard deviations seemed to be well within the manufacturer's informed precision of 0.2 °C and 1%, with values observed within 0–0.05 °C for temperature, and 0.06%–0.25% for relative humidity. These results demonstrated the HIGROTERM system as a reliable tool for the measurement of temperature and humidity, with, at least, comparable performance to previous systems used in the laboratory.

It is relevant to stress, however, that this validation study was performed as a preliminary evaluation of the performance of the HIGROTERM system, to ensure that the system presented similar values to the monitoring system used so far in the laboratory, and no attempt was made to metrologically characterize the system or to verify the manufacturer's specification. This analysis has been deemed sufficient for the uses in which HIGROTERM was employed so far within the Laboratory of Material Testing at University of Brasília; however, more strict applications may require complete characterization of the system and DHT22 sensors, with the incorporation of correction factor charts within service temperature and humidity range, full static and dynamic characterization of the sensor, and other relevant characteristics, with some studies describing calibration campaigns for temperature and humidity sensors [30,31].

3.6. Application

Since its development, the HIGROTERM system has constantly been used in the Laboratory of Material Testing at University of Brasília, mostly to monitor the temperature and humidity during material tests that are highly sensible to environmental variables and that must be performed in controlled environments, such as mortar drying and salt crystallization tests [17–20].

These tests require a climatically controlled chamber that can maintain constant temperature and humidity values throughout the tests. Commercially available equipment that guarantee the constancy of environmental variables within the variation limits stated in test standards commonly employ the active control of temperature and humidity and can be relatively expensive. To enable these tests within budget-tight research projects and teaching activities, an alternative was devised with the use of a small room within the core of the Laboratory building, for better heat and humidity insulation, with a dedicated air-conditioning system to maintain constant temperature, and sealed test containers, of various sizes, into which were positioned recipients with magnesium chloride solutions specially dosed to maintain the test chamber environment within given limits of relative humidity, which is a methodology investigated elsewhere [30,32]. In this context, the HIGROTERM system is a valuable tool for flexible monitoring temperature and humidity in various points inside test containers and around the test room, to ensure that all relevant environmental variables were within the variation limits required by the tests. With previous systems, such as the commercial system used in the validation stage of this work, this task involved many obstacles, such as high costs because various single-channel systems were necessary to monitor all locations; impossibility to monitor small containers due to bulky equipment; and inexistence of a data logging scheme that could allow later analysis of the history of measurements.

Figure 11 presents the illustration of the test setup employed in various tests performed in the Laboratory of Material Testing. Referencing Figure 11, because of the flexibility of small sensor nodes combined with cables of different lengths, the HIGROTERM system enabled monitoring of small containers, such as Nodes 1 to 4, large containers or laboratory chapels would require more than one sensor to fully represent the ambient condition, such as Nodes 6 and 7, and even points across the room, such as Nodes 5 and 8.

Figure 11. Example of test setup for material testing with the use of the HIGROTERM system.

Figure 12 presents examples of data obtained during a test campaign performed at the Laboratory of Material Testing, during an experiment of mortar drying. Figure 12a presents an example of the raw data file which is saved in the SD card in a .csv format, which can be opened with any text editor software. Figure 12b presents the data already tabulated in a spreadsheet, which can easily be obtained with the automatic formatting of .csv files widely available on spreadsheet-editing software. Figure 12c,d present the temperature and humidity history obtained during a mortar-drying test campaign, which used seven of the eight available channels of the HIGROTERM system, as shown in the graph. From Figure 12d, the differences in the humidity history between channels H2 and H5 can be observed, which were used to monitor the sealed test containers that had controlled humidity at around 50% RH due to the magnesium chloride solution, and the other channels, which were used to monitor the whole laboratory room. The room was subject to changes in humidity due to daily cycles and the entering of personnel to conduct the experiment; therefore, a higher variability could be observed in their data in comparison to H2 and H5 nodes.

It is relevant to note that in the applications used so far with the HIGROTERM system, the laboratorial environment was relatively non-aggressive to the system itself and generally non-hazardous. No hazardous agents that could interfere with its functioning or cause accidents were present in the Laboratory of Material Testing, such as excessive vibration, electromagnetic induction, and inflammable or explosive agents. In this sense, the HIGROTERM system was not tested or certified in such environments; therefore, it must be used with caution and at discretion by the user, especially in harsh environments.

3.7. Future Developments and Potentials

Some future developments and potentials can be devised from the current implementation of the HIGROTERM system. The PCB presented in this work may be further developed to include only the minimum components of the Arduino MEGA and circuitry connection of the RTC and sensor modules. By undertaking this approach, all hardware would be soldered in a single board and may be ordered already mounted from PCB manufacturers, providing scalability for the fabrication of a series of systems and miniaturization features required in some laboratorial or field applications.

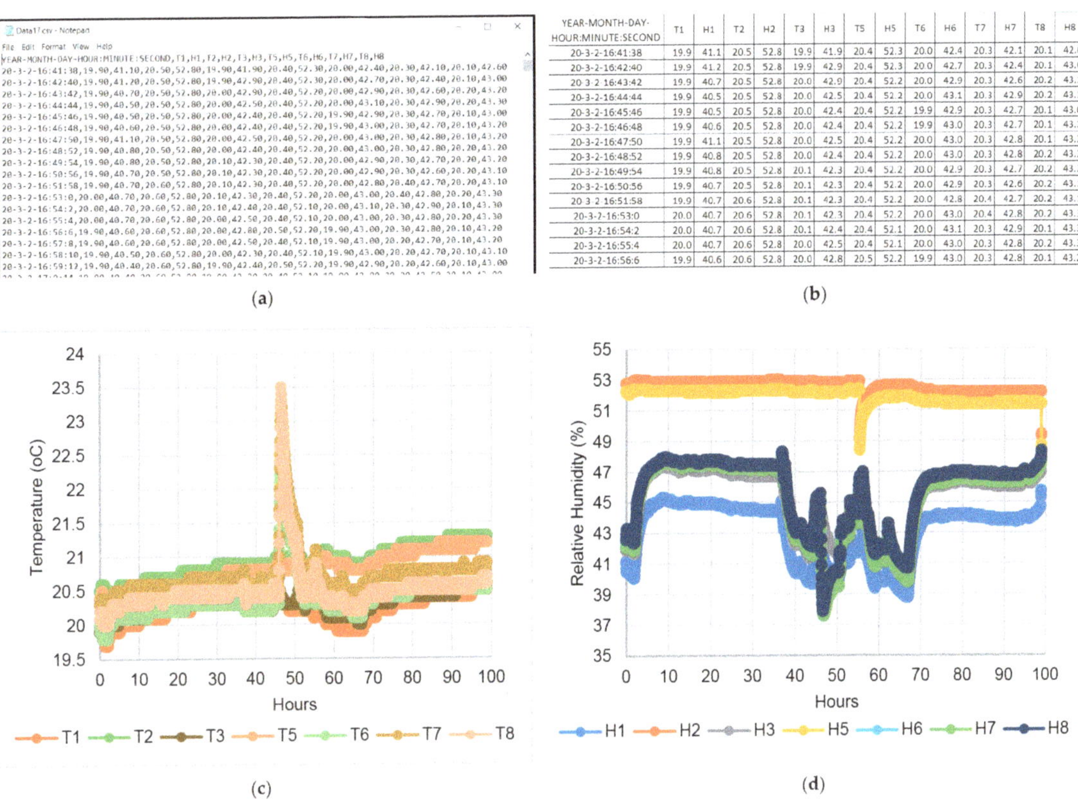

```
Data17.csv - Notepad
File Edit Format View Help
YEAR-MONTH-DAY-HOUR:MINUTE:SECOND,T1,H1,T2,H2,T3,H3,T5,H5,T6,H6,T7,H7,T8,H8
20-3-2-16:41:38,19.90,41.10,20.50,52.80,19.90,41.90,20.40,52.30,20.00,42.40,20.30,42.10,20.10,42.60
20-3-2-16:42:40,19.90,41.20,20.50,52.80,19.90,42.90,20.40,52.30,20.00,42.70,20.30,42.40,20.10,43.00
20-3-2-16:43:42,19.90,40.70,20.50,52.80,20.00,42.90,20.40,52.20,20.00,42.90,20.30,42.60,20.20,43.20
20-3-2-16:44:44,19.90,40.50,20.50,52.80,20.00,42.50,20.40,52.20,20.00,43.10,20.30,42.90,20.20,43.30
20-3-2-16:45:46,19.90,40.50,20.50,52.80,20.00,42.40,20.40,52.20,19.90,42.90,20.30,42.70,20.10,43.00
20-3-2-16:46:48,19.90,40.60,20.50,52.80,20.00,42.40,20.40,52.20,19.90,43.00,20.30,42.70,20.10,43.20
20-3-2-16:47:50,19.90,41.10,20.50,52.80,20.00,42.50,20.40,52.20,20.00,43.00,20.30,42.80,20.10,43.20
20-3-2-16:48:52,19.90,40.80,20.50,52.80,20.00,42.40,20.40,52.20,20.00,43.00,20.30,42.80,20.20,43.20
20-3-2-16:49:54,19.90,40.80,20.50,52.80,20.10,42.30,20.40,52.20,20.00,42.90,20.30,42.70,20.20,43.20
20-3-2-16:50:56,19.90,40.70,20.50,52.80,20.10,42.30,20.40,52.20,20.00,42.90,20.30,42.60,20.20,43.10
20-3-2-16:51:58,19.90,40.70,20.60,52.80,20.10,42.30,20.40,52.20,20.00,42.80,20.40,42.70,20.20,43.10
20-3-2-16:53:0,20.00,40.70,20.60,52.80,20.10,42.30,20.40,52.20,20.00,43.00,20.40,42.80,20.20,43.30
20-3-2-16:54:2,20.00,40.70,20.60,52.80,20.10,42.40,20.40,52.10,20.00,43.10,20.30,42.90,20.10,43.30
20-3-2-16:55:4,20.00,40.70,20.60,52.80,20.00,42.50,20.40,52.10,20.00,43.00,20.30,42.80,20.20,43.30
20-3-2-16:56:6,19.90,40.60,20.60,52.80,20.00,42.80,20.50,52.20,19.90,43.00,20.30,42.80,20.10,43.20
20-3-2-16:57:8,19.90,40.60,20.60,52.80,20.00,42.50,20.40,52.10,19.90,43.00,20.20,42.70,20.10,43.20
20-3-2-16:58:10,19.90,40.50,20.60,52.80,20.00,42.30,20.40,52.10,19.90,43.00,20.20,42.70,20.10,43.10
20-3-2-16:59:12,19.90,40.40,20.60,52.80,19.90,42.40,20.50,52.20,19.90,42.90,20.20,42.60,20.10,43.00
```

(a)

YEAR-MONTH-DAY-HOUR:MINUTE:SECOND	T1	H1	T2	H2	T3	H3	T5	H5	T6	H6	T7	H7	T8	H8
20-3-2-16:41:38	19.9	41.1	20.5	52.8	19.9	41.9	20.4	52.3	20.0	42.4	20.3	42.1	20.1	42.6
20-3-2-16:42:40	19.9	41.2	20.5	52.8	19.9	42.9	20.4	52.3	20.0	42.7	20.3	42.4	20.1	43.0
20-3-2-16:43:42	19.9	40.7	20.5	52.8	20.0	42.9	20.4	52.2	20.0	42.9	20.3	42.6	20.2	43.2
20-3-2-16:44:44	19.9	40.5	20.5	52.8	20.0	42.5	20.4	52.2	20.0	43.1	20.3	42.9	20.2	43.3
20-3-2-16:45:46	19.9	40.5	20.5	52.8	20.0	42.4	20.4	52.2	19.9	42.9	20.3	42.7	20.1	43.0
20-3-2-16:46:48	19.9	40.6	20.5	52.8	20.0	42.4	20.4	52.2	19.9	43.0	20.3	42.7	20.1	43.2
20-3-2-16:47:50	19.9	41.1	20.5	52.8	20.0	42.5	20.4	52.2	20.0	43.0	20.3	42.8	20.1	43.2
20-3-2-16:48:52	19.9	40.8	20.5	52.8	20.0	42.4	20.4	52.2	20.0	43.0	20.3	42.8	20.2	43.2
20-3-2-16:49:54	19.9	40.8	20.5	52.8	20.1	42.3	20.4	52.2	20.0	42.9	20.3	42.7	20.2	43.2
20-3-2-16:50:56	19.9	40.7	20.5	52.8	20.1	42.3	20.4	52.2	20.0	42.9	20.3	42.6	20.2	43.1
20-3-2-16:51:58	19.9	40.7	20.6	52.8	20.1	42.3	20.4	52.2	20.0	42.8	20.4	42.7	20.2	43.1
20-3-2-16:53:0	20.0	40.7	20.6	52.8	20.1	42.3	20.4	52.2	20.0	43.0	20.4	42.8	20.2	43.3
20-3-2-16:54:2	20.0	40.7	20.6	52.8	20.1	42.4	20.4	52.1	20.0	43.1	20.3	42.9	20.1	43.3
20-3-2-16:55:4	20.0	40.7	20.6	52.8	20.0	42.5	20.4	52.1	20.0	43.0	20.3	42.8	20.2	43.3
20-3-2-16:56:6	19.9	40.6	20.6	52.8	20.0	42.8	20.5	52.2	19.9	43.0	20.3	42.8	20.1	43.2

(b)

(c)

(d)

Figure 12. HIGROTERM results: (**a**) raw data file in .csv format; (**b**) tabulated data in after automatic formatting on a spreadsheet editor software; (**c**) temperature history during a test session (**d**) humidity history during a test session. During test sessions, 7 channels were used: T1, T2, T3, T5, T6, T7, T8.

Further capabilities may be incorporated to the HIGROTERM system. An emergency power source management to deal with power outages may be required when the data logging must not be interrupted. This could be achieved by using a battery and an appropriate circuitry that switches on when the external power source is off and recharges itself when the external power source is on. Additionally, wireless connectivity may be of interest when the user desires to remotely monitor the data logging in real time, and solutions for this using the Arduino platform or similar boards can be found in the literature [9,12,13,15]. The HIGROTERM source code and hardware may also serve as a basis for systems with other objectives, such as environment control systems to simulate a given temperature and humidity condition or history. This can be performed by associating the sensor measurements to the control of power loads such as heaters, coolers, humidifiers, and dehumidifiers.

The number of sensor channels may also be easily increased because, based on the presented implementation, there are still 4 digital pins and 14 analog pins available in the Arduino MEGA. The DHT22 requires just one digital or analog pin to communicate with the MCU and power and ground lines, while presenting a maximum current drain of 1.5 mA [16]. In this way, all available digital and analog pins may be used as sensor channels. However, such modification also implies modifying the system source code, so the larger number of sensor channels can be properly handled, both on the channel configuration, data storing, and LCD displaying routines.

Although the HIGROTERM system was designed using the Arduino MEGA board, the hardware described, and the system source code are compatible with other boards. Successful tests have been performed with the Arduino DUE board, which is a board with similar layout but with better specifications in comparison to MEGA. The main restriction is the size of the sketch, which currently occupies 131,232 kB of Flash memory. For other boards, the number of pins required to connect the LCD touch screen, RTC and sensors must also be considered.

As already stated, no tests for safety certification were conducted with the HIGROTERM, which is, therefore, not considered an industrial equipment. To allow more general uses of the system, it is necessary to perform the required tests to ensure that the system meets the appropriate safety requirements.

4. Conclusions

This paper presented the HIGROTERM system, an open-source, low-cost, multi-channel data-acquisition system for temperature and humidity measurement and logging, with an easy-to-use interface based on an LCD touchscreen developed with the Arduino prototyping platform. Comprehensive details about the system, including its design, source-code logic, hardware assemblage, bill of materials, and capabilities were provided to allow the interested reader to fully replicate the system or even further develop the current system to fine-tune it to particular tasks or add useful functionalities. Validation results and an application example were also presented, so the readers can acquire a preliminary notion of the system accuracy and precision, and devise ways in which the system may be useful for their own activities. The authors expect that the HIGROTERM system may be especially useful in budget-tight researching and teaching activities and, from its replication by the community, may be a source of innovation and interest in low-cost electronics for real problem solving in various fields of science.

Author Contributions: Conceptualization, R.R.R. and E.B.; methodology, R.R.R. and R.L.; software, R.R.R.; validation, R.R.R.; formal analysis, R.R.R.; investigation, R.R.R.; resources, E.B.; data curation, R.R.R.; writing—original draft preparation, R.R.R.; writing—review and editing, R.R.R. and R.L.; visualization, R.R.R.; supervision, E.B. and R.L.; project administration, R.R.R. and E.B.; funding acquisition, E.B. All authors have read and agreed to the published version of the manuscript.

Funding: This research was funded by Conselho Nacional de Desenvolvimento Científico e Tecnológico, grant number 158274/2019-9.

Institutional Review Board Statement: Not applicable.

Informed Consent Statement: Not applicable.

Data Availability Statement: The data presented in this study are openly available in GitHub at https://doi.org/10.5281/zenodo.5525381 DOI.

Conflicts of Interest: The authors declare no conflict of interest.

References

1. Arduino What is Arduino?/Why Arduino? Available online: https://www.arduino.cc/en/Guide/Introduction (accessed on 30 April 2020).
2. Mesas-Carrascosa, F.J.; Verdú Santano, D.; Meroño, J.E.; Sánchez de la Orden, M.; García-Ferrer, A. Open source hardware to monitor environmental parameters in precision agriculture. *Biosyst. Eng.* **2015**, *137*, 73–83. [CrossRef]
3. Cao Pham, T.; Bich Vo, H.; Quang Tran, N. A design of greenhouse monitoring system based on low-cost mesh Wi-Fi wireless sensor network: *note: Sub-titles are not captured in Xplore and should not be used. In Proceedings of the 2021 IEEE International IOT, Electronics and Mechatronics Conference (IEMTRONICS), Toronto, ON, Canada, 21–24 April 2021; pp. 2–7. [CrossRef]
4. Ichwana; Nasution, I.S.; Sundari, S.; Rifky, N. Data Acquisition of Multiple Sensors in Greenhouse Using Arduino Platform. In *IOP Conference Series: Earth and Environmental Science*; IOP Publishing: West Sumatera, Indonesia, 2020; Volume 515. [CrossRef]
5. Gaikwad, S.V.; Vibhute, A.D.; Kale, K.V.; Mehrotra, S.C. An innovative IoT based system for precision farming. *Comput. Electron. Agric.* **2021**, *187*, 106291. [CrossRef]
6. Chu, M.; Patton, A.; Roering, J.; Siebert, C.; Selker, J.; Walter, C.; Udell, C. SitkaNet: A low-cost, distributed sensor network for landslide monitoring and study. *HardwareX* **2021**, *9*, e00191. [CrossRef]

7. Winkler, R. MeteoMex: Open infrastructure for networked environmental monitoring and agriculture 4.0. *PeerJ Comput. Sci.* **2021**, *7*, e343. [CrossRef] [PubMed]
8. Ali, A.S.; Zanzinger, Z.; Debose, D.; Stephens, B. Open Source Building Science Sensors (OSBSS): A low-cost Arduino-based platform for long-term indoor environmental data collection. *Build. Environ.* **2016**, *100*, 114–126. [CrossRef]
9. Karami, M.; McMorrow, G.V.; Wang, L. Continuous monitoring of indoor environmental quality using an Arduino-based data acquisition system. *J. Build. Eng.* **2018**, *19*, 412–419. [CrossRef]
10. Asinelli, M.G.; Serra, M.S.; Marimòn, J.M.; Espaulella, J.S. The smARTS_Museum_V1: An open hardware device for remote monitoring of Cultural Heritage indoor environments. *HardwareX* **2018**, *4*, e00028. [CrossRef]
11. Turhan, C.; Simani, S.; Gokcen Akkurt, G. Development of a personalized thermal comfort driven controller for HVAC systems. *Energy* **2021**, *237*, 121568. [CrossRef]
12. Zafar, S.; Miraj, G.; Baloch, R.; Murtaza, D.; Arshad, K. An IoT Based Real-Time Environmental Monitoring System Using Arduino and Cloud Service. *Eng. Technol. Appl. Sci. Res.* **2018**, *8*, 3238–3242. [CrossRef]
13. Utepov, Y.; Khudaibergenov, O.; Kabdush, Y.; Kazkeev, A. Prototyping an embedded wireless sensor for monitoring reinforced concrete structures. *Comput. Concr.* **2019**, *24*, 95–102. [CrossRef]
14. Fathihah, M.A.; Khairunnisa, M.P.; Rashid, M.; Norruwaida, J.; Dewika, M.; Ito, Y.; Wuled Lenggoro, I. Development of low-cost and user-friendly sustainable portable particulate sensor. In Proceedings of the IOP Conference Series: Materials Science and Engineering, Kuala Lumpur, Malaysia, 13–14 August 2018; Volume 458. [CrossRef]
15. Courtemanche, J.; King, S.; Bouck, D. Engineering Novel Lab Devices Using 3D Printing and Microcontrollers. *SLAS Technol.* **2018**, *23*, 448–455. [CrossRef] [PubMed]
16. Liu, T. Digital-output relative humidity & temperature sensor/module DHT22. *Aosong Electron.* **2011**.
17. Rilem, T.C. RILEM TC 127-MS MS-A.2-uni-directional salt crystallization test for masonry units. *Mater. Struct.* **1998**, *31*, 10–11.
18. Bekker, P.; Borchelt, G.; Bright, N.; Emrich, F.; Forde, M.; Gallegos, H.; Groot, C.; Hedstrom, E.; Lawrence, S.; Maurenbrecher, P.; et al. RILEM TC 127-Ms: Tests for Masonry Materials and Structures. *Mater. Struct. Constr.* **1998**, *31*, 2–19.
19. Lubelli, B.; Cnudde, V.; Diaz-Goncalves, T.; Franzoni, E.; van Hees, R.P.J.; Ioannou, I.; Menendez, B.; Nunes, C.; Siedel, H.; Stefanidou, M.; et al. Towards a more effective and reliable salt crystallization test for porous building materials: State of the art. *Mater. Struct.* **2018**, *51*, 55. [CrossRef]
20. European Standards. *For EN 16322: Conservation of Cultural Heritage-Test Methods-Determination of Drying Properties*; CSN: Brussels, Belgium, 2013.
21. Arduino Arduino Mega 2560 Rev3. Available online: https://store.arduino.cc/usa/mega-2560-r3 (accessed on 30 April 2020).
22. Atmel. Atmel ATmega640/V-1280/V-1281/V-2560/V2561/V—Datasheet 2014. *Atmel.* 2014. Available online: https://ww1.microchip.com/downloads/en/devicedoc/atmel-2549-8-bit-avr-microcontroller-atmega640-1280-1281-2560-2561_datasheet.pdf (accessed on 23 September 2021).
23. Arduino Shields. Available online: https://www.arduino.cc/en/Main/arduinoShields (accessed on 30 April 2020).
24. Rocha Ribeiro, R. renr3/HIGROTERM: First release (v0.1). Zenodo. 2021. Available online: https://zenodo.org/record/5525381 (accessed on 10 October 2021).
25. Open Source Initiative The MIT License. Available online: https://opensource.org/licenses/MIT (accessed on 11 June 2020).
26. Creative Commons Attribution-NonCommercial-ShareAlike 4.0 International (CC BY-NC-SA 4.0). Available online: https://creativecommons.org/licenses/by-nc-sa/4.0/ (accessed on 1 May 2020).
27. Maxim Integrated. DS 3231 RTC General Description. *Maxim Integrated.* 2015. Available online: https://datasheets.maximintegrated.com/en/ds/DS3231.pdf (accessed on 23 September 2021).
28. Arduino Arduino IDE. Available online: https://www.arduino.cc/en/software (accessed on 16 July 2021).
29. Kasvi Termômetros. Available online: https://kasvi.com.br/wp-content/uploads/2015/04/Termometros.pdf (accessed on 3 September 2021).
30. Fossa, M.; Petagna, P. Use and calibration of capacitive RH sensors for the hygrometric control of the CMS tracker. *CERN-CMS-NOTE* **2003**, *24*, 1–11.
31. Koestoer, R.A.; Pancasaputra, N.; Roihan, I.; Harinaldi. A simple calibration methods of relative humidity sensor DHT22 for tropical climates based on Arduino data acquisition system. In Proceedings of the AIP Conference Proceedings, Bali, Indonesia, 16–17 November 2018; Volume 2062.
32. Quincot, G.; Azenha, M.; Barros, J.; Faria, R. Use of salt solutions for assuring constant relative humidity conditions in contained environments. *Univ. Minho Guimaraes Port.* **2011**, *1*, 1–32.

Article

A New No Equilibrium Fractional Order Chaotic System, Dynamical Investigation, Synchronization, and Its Digital Implementation

Zain-Aldeen S. A. Rahman [1,2], Basil H. Jasim [2], Yasir I. A. Al-Yasir [3,*], Raed A. Abd-Alhameed [3] and Bilal Naji Alhasnawi [2]

1 Department of Electrical Techniques Institute/Qurna, Southern Technical University, Basra 61016, Iraq; as.zain9391@stu.edu.iq
2 Department of Electrical Engineering, College of Engineering, University of Basrah, Basra 61004, Iraq; basil.jasim@uobasrah.edu.iq (B.H.J.); bilalnaji11@yahoo.com (B.N.A.)
3 Biomedical and Electronics Engineering, Faculty of Engineering and Informatics, University of Bradford, Bradford BD7 1DP, UK; R.A.A.Abd@bradford.ac.uk
* Correspondence: y.i.a.al-yasir@bradford.ac.uk; Tel.: +44-127-423-8047

Abstract: In this paper, a new fractional order chaotic system without equilibrium is proposed, analytically and numerically investigated, and numerically and experimentally tested. The analytical and numerical investigations were used to describe the system's dynamical behaviors including the system equilibria, the chaotic attractors, the bifurcation diagrams, and the Lyapunov exponents. Based on the obtained dynamical behaviors, the system can excite hidden chaotic attractors since it has no equilibrium. Then, a synchronization mechanism based on the adaptive control theory was developed between two identical new systems (master and slave). The adaptive control laws are derived based on synchronization error dynamics of the state variables for the master and slave. Consequently, the update laws of the slave parameters are obtained, where the slave parameters are assumed to be uncertain and are estimated corresponding to the master parameters by the synchronization process. Furthermore, Arduino Due boards were used to implement the proposed system in order to demonstrate its practicality in real-world applications. The simulation experimental results were obtained by MATLAB and the Arduino Due boards, respectively, with a good consistency between the simulation results and the experimental results, indicating that the new fractional order chaotic system is capable of being employed in real-world applications.

Keywords: fractional order; dynamics; chaotic; system; synchronization; arduino due

Citation: Rahman, Z.-A.S.A.; Jasim, B.H.; Al-Yasir, Y.I.A.; Abd-Alhameed, R.A.; Alhasnawi, B.N. A New No Equilibrium Fractional Order Chaotic System, Dynamical Investigation, Synchronization, and Its Digital Implementation. *Inventions* **2021**, *6*, 49. https://doi.org/10.3390/inventions6030049

Academic Editors: Francisco Manzano Agugliaro and Esther Salmerón-Manzano

Received: 13 June 2021
Accepted: 30 June 2021
Published: 6 July 2021

1. Introduction

Research on chaotic systems has had a significant practical effect since Lorenz established chaos theory in 1963 [1]. Over the last few decades, nonlinear phenomena in chaos have been widely employed in engineering, science, and applied mathematics [2–4]. Chaos systems with hidden attractors have been the focus of recent research. Self-excited attractors and hidden attractors have been classed as chaotic attractors in dynamical systems since the seminal article by Leonov et al. was investigated [5]. The unstable equilibrium points (system equilibria) are responsible for exciting the basin of attraction in the self-excited chaotic [6]. An attractor, on the other hand, is said to be hidden if its basin of attraction does not intersect with any of the small neighborhoods of the unstable equilibrium [7]. Hidden chaotic attractors are chaotic attractors in dynamical systems with stable equilibria, no equilibrium, and surfaces or lines of equilibria [8].

Chaotic complex systems with hidden attractors are vital in a wide range of scientific and engineering fields, such as bridge wings design [9], induction motors for drilling [10], chemical reactors systems [11], aircraft control systems [12], memristive circuits [13], encryption [14], PLL systems [15], weather prediction systems [16], and secure communication schemes [17].

Hidden attractors research in the past has primarily concentrated on integer-order dynamical systems. There have been several studies on complicated chaotic systems with hidden chaotic attractors, such as in [18–23]. The fractional order derivative and fraction order integration calculus have recently received a lot of attention, owing to the fractional calculus providing more accurate models than the integer order [24].

Fractional calculus can be thought of as an expansion of traditional calculus as a branch of mathematical analysis. Due to its possible applications in a variety of domains, the study of fractional calculus has recently received a lot of attention [25]. Fractional calculus can describe many systems in transdisciplinary domains. Furthermore, the fractional-order model can provide an explicit description of the physical process and provide additional insight [26]. Fractional calculus can be used in control, bio-engineering, oscillators, analog filters, circuit theory, image processing encryption systems, and chemistry [27]. Fractional order chaotic models have a more complex dynamical behavior than integer models (since they include the fractional order parameter as well as the original system characteristics); as a result, they are important in secure communications systems [28,29].

The synchronization technique of chaos is based on the principle that two chaotic systems may develop on different attractors, but when synchronized, they begin on different attractors and later follow a single trajectory. When the trajectories of two systems are matched, this synchronization is achieved between these two systems [30]. The control and synchronization techniques of the fractional order chaotic systems can be considered the fundamental challenge of using these systems in many applications as robotics, cryptography, mechanical, and secure communication applications [31,32]. To control and synchronize the fractional-order chaotic systems, a variety of control and synchronization techniques have been developed, such as active control [33], sliding mode control [34], adaptive control [35], passive control [36], and impulsive control [37].

We suggested a new 3D fractional order chaotic system with no equilibrium in this paper, so it can excite hidden chaotic attractors. The system dynamical behaviors including the system equilibria, the chaotic attractors, the bifurcation diagrams, and the Lyapunov exponents were analytically and numerically investigated. Then, two identical new systems, one working as the master (drive) and the other as the slave (response), were used to develop a synchronization mechanism based on adaptive control theory.

Based on Lyapunov's stability theory, the adaptive control laws responsible for tracking and aligning the slave state variables (slave trajectories) with the equivalent trajectories in the master side. Consequently, the update laws for updating the uncertain slave parameters were obtained. By this scenario, the slave was well synchronized with the master. The MATLAB was used to verify our work, testing, and results. Additionally, Arduino Due boards were used to implement a workable hardware electronic circuit for the new system.

The remainder of this paper is organized as follows: the basic mathematical background of the fractional order systems is introduced in Section 2. Section 3 introduces a new fractional order chaotic system and determines its chaotic attractors classes, as well as the system equilibria. In Section 4, the suggested system's dynamical behavior properties are investigated using Lyapunov exponents and bifurcation diagrams. In Section 5, we establish a synchronization approach between two identical new systems, using Lyapunov stability to drive the adaptive control laws and the update laws to achieve the synchronization mechanism and estimate the uncertain slave parameters, respectively. In Section 6, we use the Arduino Due boards to implement the real electronic circuit for the suggested system. In Section 7, we conclude this paper.

2. Fundamentals of Fractional Order Systems

Integer-order calculus is mathematically extended to fractional calculus. While it offers the advantages of integer order calculus, it also has its own logic and laws. In the definitions of fractional calculus, the Caputo, Riemann–Liouville (RL), and Grunwald–Letnikov (GL) concepts are commonly used [38].

The Gamma function noted by $\Gamma(.)$, which is defined in Equation (1), is the basic function used in fractional order calculus [39]:

$$\Gamma(n) = \int_{0}^{+\infty} e^{-t} t^{n-1} dt \ ; \quad n > 0 \tag{1}$$

where, $\Gamma(1) = 1$, $\Gamma(0) = +\infty$.

The fractional order calculus introduced by Caputo is stated as follows in Equation (2) [40]:

$$ {}_{t_0}D_t^q f(t) = \begin{cases} \frac{1}{\Gamma(k-q)} \int_{t_0}^{t} \frac{f^{(k)}(\tau)}{(t-\tau)^{q-k+1}} d\tau; & k-1 < q < k \\ \frac{d^k f(t)}{dt^k} \ ; & q = k \end{cases} \tag{2}$$

where k is the first integer number that is greater than or equal to fractional q-order.

Equation (3) provides the fractional integral operator (J^q) established by Riemann–Liouville for order ($q \geq 0$) of function ($f(t)$) [41].

$$J^q f(t) = \begin{cases} \frac{1}{\Gamma(q)} \int_{0}^{t} (t-\tau)^{q-1} f(t) d\tau \ ; & q < 0 \\ f(t) \ ; & q = 0 \end{cases} \tag{3}$$

As shown in Equation (4), Grunwald–Letnikov approaches the fractional derivative [42]:

$$D^q x(t) = f(x,t) = \lim_{h \to 0} h^{-q} (-1) \sum_{j=0}^{t/h} (-1) \begin{pmatrix} q \\ j \end{pmatrix} x(t - jh) \tag{4}$$

where, h is the step size.

3. A New Fractional Order Chaotic Model

Fractional order chaotic systems are a category of nonlinear systems that, in addition to the fundamental characteristics of integer order chaotic systems, have extra features such as extreme complexity and severe sadness behavior [43]. The following Equation (5) can be used to describe the mathematical model of the new three-dimensional fractional order chaotic system proposed in this paper:

$$\frac{d^q x}{dt^q} = az + xy; \ \frac{d^q y}{dt^q} = b - x^2; \ \frac{d^q z}{dt^q} = -cx \tag{5}$$

where the state variables are x, y, and z; the positive constant system parameters are a, b, and c, and q presents the fractional order ($0 < q < 1$).

The system in Equation (5) exhibits chaotic behavior throughout a wide range of a, b, and c parameter values, as well as fractional order (q). The system parameters for the numerical simulation are $a = 0.5$, $b = 1.8$, $c = 8$, and $q = 0.99$, with initial conditions $(x(0), y(0), z(0)) = (1, 1, 1)$. In 2017, Roberto Garrappa [44] introduced a method for solving the nonlinear fractional order systems, where the simulation results of system (Equation (5)) are obtained based on Roberto Garrappa's method with step size ($h = 0.005$). As illustrated in Figures 1 and 2, the relevant time series of the system states and phase portraits as projections on various planes are obtained, respectively.

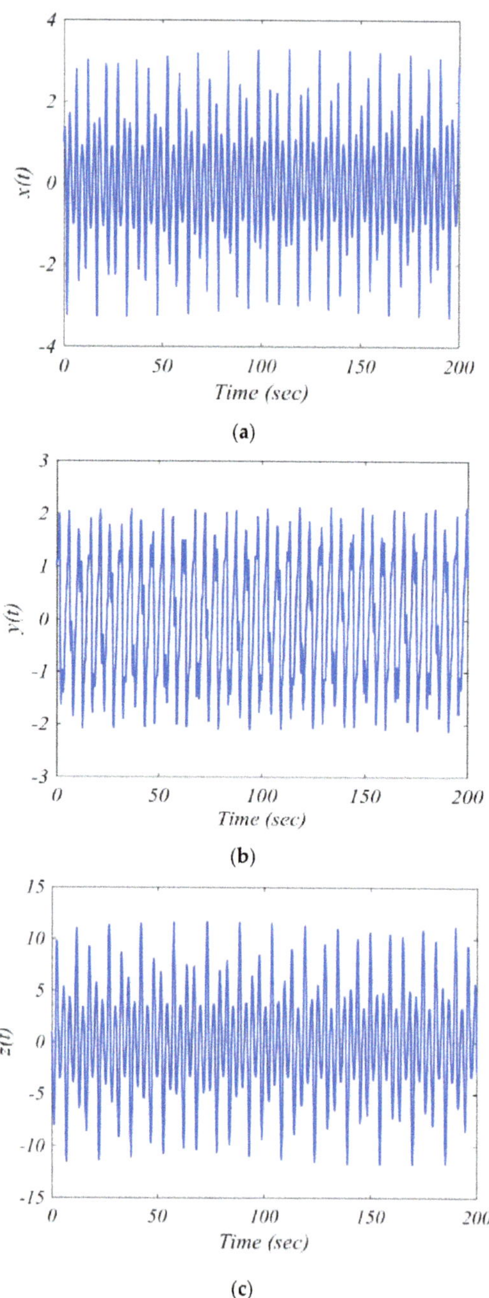

Figure 1. States time series of proposed system; (**a**) $x(t)$; (**b**) $y(t)$; (**c**) $z(t)$.

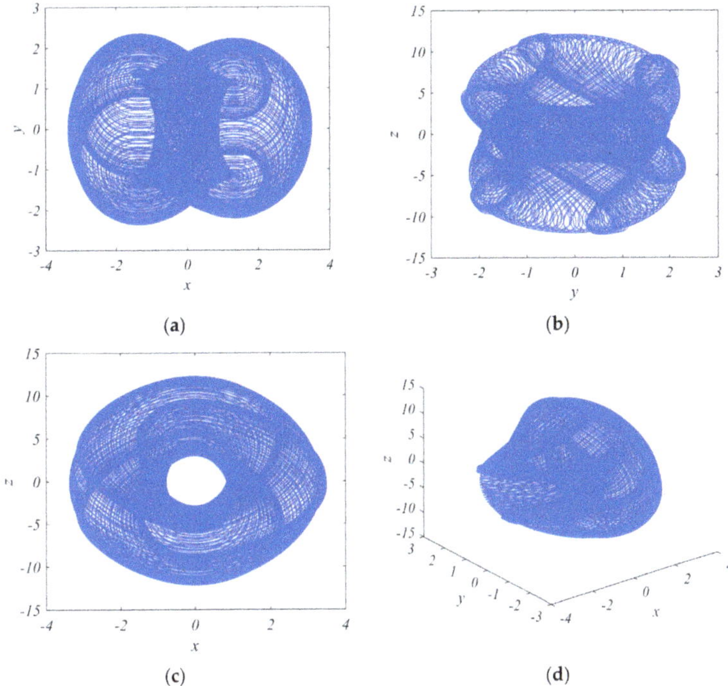

Figure 2. System (Equation (5)) phase portraits: (a) x-y chaotic attractor; (b) y-z chaotic attractor; (c) x-z chaotic attractor; (d) three-dimensional chaotic attractor.

For determining the equilibrium points (equilibria) of the proposed fractional order chaotic system the state of Equation (5) is equaled by zero as follows in Equation (6):

$$\frac{d^q x}{dt^q} = az + xy = 0; \ \frac{d^q y}{dt^q} = b - x^2 = 0; \ \frac{d^q z}{dt^q} = -cx = 0 \quad (6)$$

As can be noted in Equation (6), there is an inconsistency in Equation (6), where the state variable $x = 0$ can be obtained from the third term in Equation (6); however, it is not possible to solve the second term in the same Equation (6) because the constant b cannot equal zero. Therefore, Equation (6) has no solution, which leaves the proposed fractional order chaotic system in Equation (5) without equilibrium points, i.e., the new fractional order chaotic system can excite hidden chaotic attractors.

4. The System Dynamical Analyses

Generally, the bifurcation diagrams and Lyapunov exponents are the two main dynamical tools that can be used to investigate the dynamical behaviors of nonlinear chaotic systems [45]. In this section, the bifurcation diagrams and the Lyapunov exponents are numerically investigated by using MATLAB.

4.1. Bifurcation Diagrams

The bifurcation diagrams are important means in the nonlinear dynamics and chaos theory. In this work, for exploring the system's dynamical behavior by the bifurcation diagrams, the state variable $y(t)$ of the suggested system is plotted in contradiction with the system parameter b, and with respect to the system fractional order (q).

The influence of the system parameter a on the system dynamical behavior are obtained by the bifurcation diagrams as illustrated in Figure 3, where Roberto Garrappa's method with step size ($h = 0.005$) and an original program we designed were used for plotting the bifurcation diagrams. The other system parameters are chosen as $a = 0.5$ and $c = 8$, with initial conditions $(x(0), y(0), z(0)) = (1, 1, 1)$ and fractional order ($q = 0.98$). As can be seen in Figure 3, the new system exhibits chaotic behavior when the parameter $b > 0$.

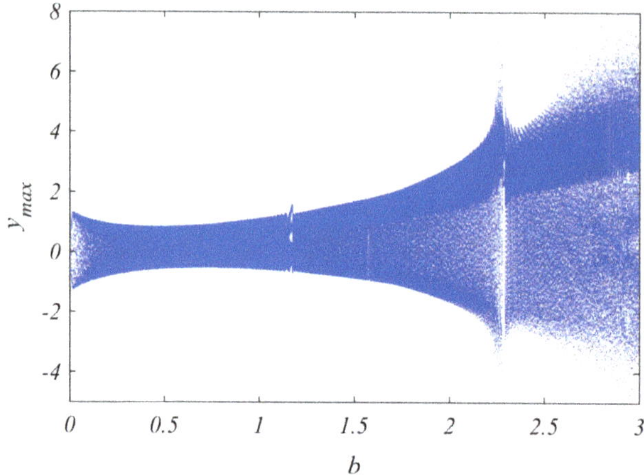

Figure 3. The bifurcation diagram verse system parameter (b).

We then choose the fractional order (q) to be the bifurcation parameter and fixed the parameters $a = 0.5$, $b = 1.8$, and $c = 8$ with the initial conditions (1, 1, 1); the dynamical behavior of the system shown in Equation (5) is obtained by the bifurcation diagrams as shown in Figure 4. As can be noted from the bifurcation diagram in Figure 4, the chaotic behavior is excited when the system fractional order $q > 0.965$. It can be seen from Figures 3 and 4, that the system in Equation (5) displays various bifurcation topological patterns.

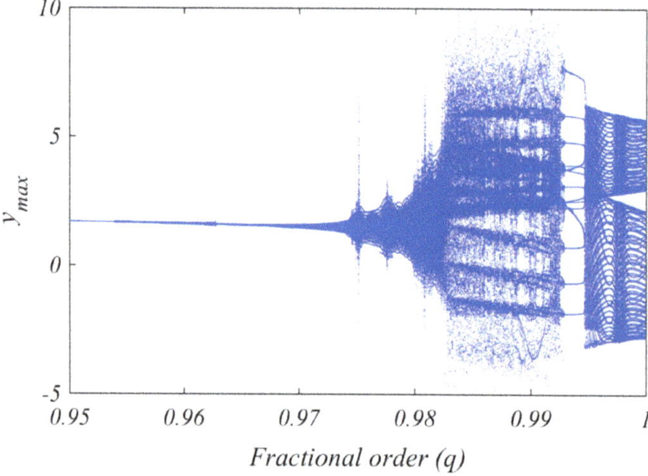

Figure 4. The bifurcation diagram verses the system fractional order (q).

This result shows that the selected system fractional order determines the generation of the hidden chaotic attractor.

4.2. Lyapunov Exponents

The Lyapunov exponents are calculated to strongly indicate that the new system exhibits the chaoticity phenomenon; at least one positive Lyapunov exponent in the nonlinear dynamics system ensures these systems exhibit chaos [46].

Figure 5 demonstrates the Lyapunov exponents with respect to the time (1000 s), where the system parameters are chosen as $a = 0.5$, $b = 1.8$, and $c = 8$, with initial conditions ($x(0)$, $y(0)$, $z(0)$) = (1, 1, 1) and fractional order ($q = 0.98$). The consistent Lyapunov exponents are obtained as Le1 = 0.2384, Le2 = −0.2859, and Le3 = −0.3681. The existence of the positive Lyapunov exponent (Le3) is enough to prove the system in Equation (5) can exhibit the chaos. It should be noted that because Le1 + Le2 + Le3 = −0.3256, the system in Equation (5) is dissipative, i.e., the new system state trajectories converge into a weird attractor upon conclusion.

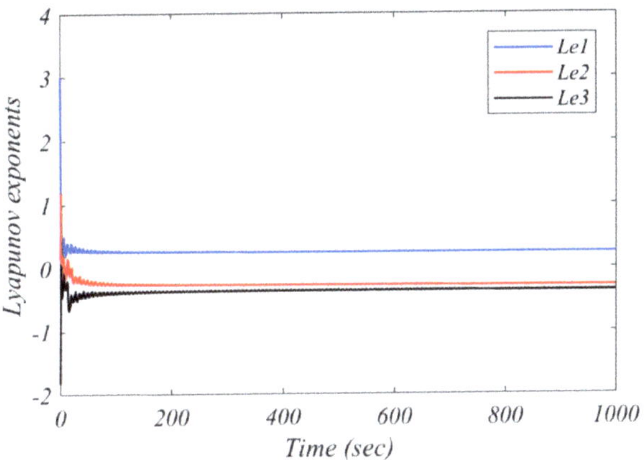

Figure 5. Lyapunov exponents corresponding to time.

In addition, Lyapunov exponents are calculated with respect to changing the fractional order to $q \in [0.95,1]$ as shown in Figure 6. The used system parameters are as mentioned in the first Lyapunov exponents calculations. As can be seen in Figure 6, the Lyapunov exponents are Le1 = 0.0287, Le2 = 0.0019, and Le3 = −0.0017, which clearly indicate chaotic attractors in the proposed system.

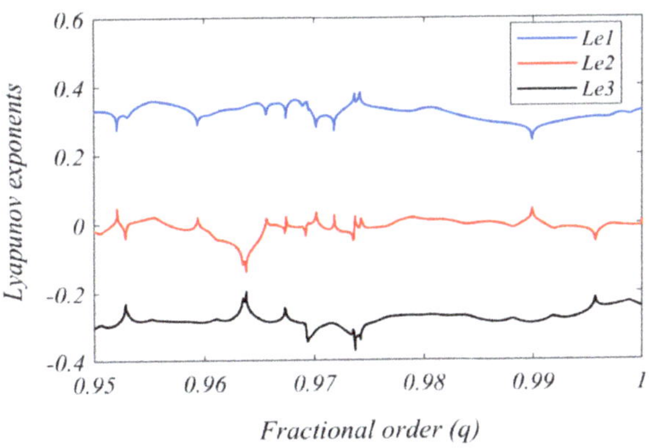

Figure 6. Lyapunov exponents corresponding to the system in Equation (5) fractional order (q).

5. Adaptive Synchronization of Two New Fractional Order Chaotic Systems

Particularly, because the fractional order chaotic complex systems have an extra complexity in dynamical behavior and cannot be described by classical mathematical methods (i.e., deterministic systems), these systems are capable of being widely used in secure communication, image processing, and cryptography systems [47]. Therefore, the fractional order chaotic system synchronization mechanisms have received much attention due to the importance of applications that can be extended into many fields such as physics, engineering, computer science, biology, economics, and brain science. The drive-response (master-slave) form is considered the basic configuration of the chaos synchronization mechanism, where the trajectories of the slave chaotic system must track the trajectories of the master chaotic system. Several methods have been developed to attain the chaos synchronization in fractional-order chaotic systems as mentioned in Section 1.

In this work, we developed an adaptive synchronization technique in order to achieve synchronization between two identical new systems, where the first acts as the master (drive) system and the second acts as the slave (response) system. Based on Lyapunov's theory, the synchronization controllers and the slave parameter estimation laws were derived. The adaptive control laws drive the slave so its trajectories and track the analogues master trajectories, and the slave parameter estimation laws are used for updating the uncertain slave parameters corresponding to the known analogous master parameters.

The adaptive synchronization technique has many benefits, such as good transient performance, rapid dynamics responses, and a robust system for parameter variations and initial conditions.

5.1. Adaptive Controller Design Process

In this subsection, we present the design of the adaptive controller for achieving the synchronization between two identical new fractional order chaotic systems. The master and slave state equations are presented by Equations (7) and (8), respectively:

$$\frac{d^q x_m}{dt^q} = a_m z_m + x_m y_m; \ \frac{d^q y_m}{dt^q} = b_m - {x_m}^2; \ \frac{d^q z}{dt^q} = -c_m x_m \quad (7)$$

$$\frac{d^q x_s}{dt^q} = a_s(t) z_s + x_s y_s + u_1; \ \frac{d^q y_s}{dt^q} = b_s(t) - {x_s}^2 + u_2; \ \frac{d^q z}{dt^q} = -c_s(t) x_s + u_3 \quad (8)$$

where, u_1, u_2, and u_3 are the adaptive synchronization controllers that we want to design; a_m, b_m, and c_m are the known master parameters; and $a_s(t)$, $b_s(t)$, and $c_s(t)$ present the uncertain slave parameters that must be estimated. The master-slave synchronization errors can be determined by Equation (9).

$$e_x = x_s - x_m; \ e_y = y_s - y_m; \ e_z = z_s - z_m \quad (9)$$

The dynamic errors are determined as in Equation (10).

$$\frac{d^q e_x}{dt^q} = a_2 e_z + x_s y_s - x_m y_m + z_m e_a + u_1; \ \frac{d^q e_y}{dt^q} = e_b - {x_s}^2 + {x_m}^2 + u_2; \ \frac{d^q e_z}{dt^q} = c_2 e_x + x_m e_c + u_3 \quad (10)$$

where e_a, e_b, and e_c are the errors of master-slave parameters and can be determined as in Equation (11).

$$e_a = a_s(t) - a_m; \ \ e_b = b_s(t) - b_m; \ e_c = c_s(t) - c_m \quad (11)$$

The dynamics of the parameter errors can be calculated by Equation (12).

$$\dot{e}_a = \dot{a}_s(t); \ \dot{e}_b = \dot{b}_s(t); \ \dot{e}_c = \dot{c}_s(t) \quad (12)$$

We used the Lyapunov strategy to design the adaptive controllers to verify the adaptive master-slave synchronization mechanism. Consequently, the update laws for estimating the uncertain parameters were obtained. Therefore, the quadratic positive Lyapunov functions are used as in Equation (13).

$$V(e_x, e_y, e_z, e_a, e_b, e_c) = \frac{1}{2}\left(e_x^2 + e_y^2 + e_z^2 + e_a^2 + e_b^2 + e_c^2\right) \tag{13}$$

The results of the Lyapunov function dynamics are obtained as in Equation (14).

$$\dot{V} = (e_x\frac{d^q e_x}{dt^q} + e_y\frac{d^q e_y}{dt^q} + e_z\frac{d^q e_z}{dt^q} + e_a\dot{e}_a + e_b\dot{e}_b + e_c\dot{e}_c) \tag{14}$$

Equations (10)–(12) are substituted in Equation (14) to obtain the following Equation (15).

$$\begin{aligned}\dot{V} = \; & (e_x(a_2e_z + x_sy_s - x_my_m + z_me_a + u_1) + e_y(e_b - {x_s}^2 + {x_m}^2 + u_2) \\ & + e_z(c_2e_x + x_me_c + u_3) + (a_s(t) - a_m)\dot{a}_s(t) + (b_s(t) \\ & - b_m)\dot{b}_s(t) + (c_s(t) - c_m)\dot{c}_s(t)\end{aligned} \tag{15}$$

Thus, Equation (15) was used to design the synchronization controllers as in the following Equation (16).

$$u_1 = -k_xe_x - a_se_z - y_sz_s + y_mz_m \; ; \; u_2 = -k_ye_y + {x_s}^2 - {x_m}^2 \; ; \; u_3 = -k_ze_z - c_se_x \tag{16}$$

In Equation (16), k_x, k_y, and k_z present positive constants, and the uncertain slave system parameters included (a_s, b_s, and c_s) are estimated by updated laws as in the following Equation (17).

$$\dot{a}_s(t) = -z_me_x; \; \dot{b}_s(t) = -e_y; \; \dot{c}_s(t) = -x_me_z \tag{17}$$

Finally, the following Equation (18) is obtained by substituting Equations (16) and (17) in Equation (15).

$$\dot{V} = -k_x{e_x}^2 - k_y{e_y}^2 - k_zy^2 \tag{18}$$

As can be seen from Equation (17), there is a negative definite function [48]. As a result, given any initial conditions, the synchronization state errors and the master-slave parameter estimation error converge to zero exponentially with respect to time.

5.2. Simulation Results

Numerical studies using the MATLAB platform are used to confirm the efficiency of the suggested synchronization strategy. Table 1 shows the values of the master and slave parameters, fractional orders, and initial conditions utilized to simulate the aforementioned synchronization technique.

Table 1. Master and slave parameters values.

Master System		Slave System	
Parameter	**Value**	**Parameter**	**Value**
a_m	0.5	$a_s(t)$	Estimated
b_m	1.8	$b_s(t)$	Estimated
c_m	8	$c_m(t)$	Estimated
fractional order (q)	0.98	fractional order (q)	0.98
$X_m(0)$	1	$X_s(0)$	2
$y_m(0)$	1	$y_s(0)$	0
$z_m(0)$	1	$z_s(0)$	−1

The slave trajectories (state variables) effectively follow the master trajectories based on the derived adaptive control laws in Equation (16) as shown in Figure 7. Although initial conditions have different signs and values, the simulation results demonstrate that the master and slave state variables were synchronized in a short time, indicating that the developed controller is efficient. In Figure 8, the synchronization errors (e_x, e_y, and e_z) are illustrated. It can be seen that the synchronization errors rapidly (in less than 1.5 s) decrease to zero values exponentially. As illustrated in Figure 9, these uncertain slave parameters were appropriately estimated to the corresponding master parameters using the updated laws in Equation (17).

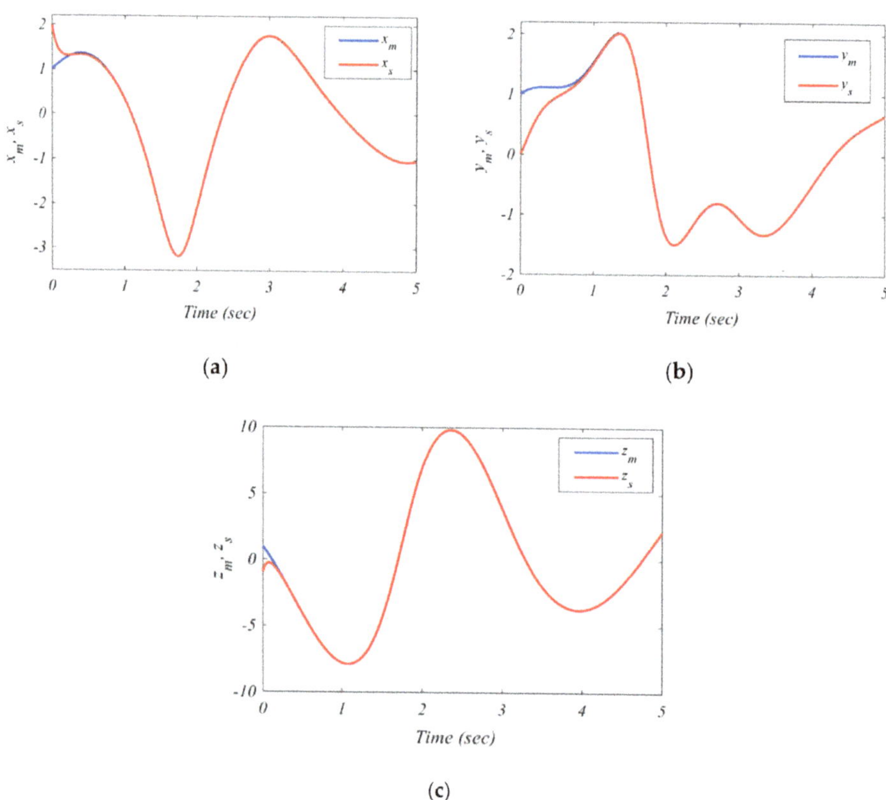

Figure 7. Tracking the slave trajectories corresponding to master trajectories; (**a**) x_s track x_m; (**b**) y_s track y_m; (**c**) z_s track z_m.

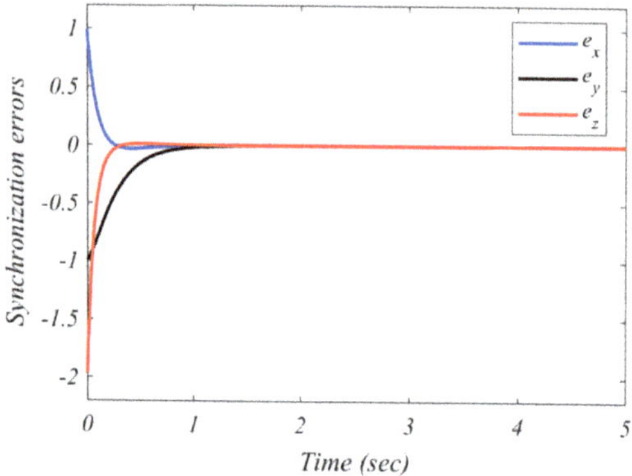

Figure 8. The state variables synchronization errors.

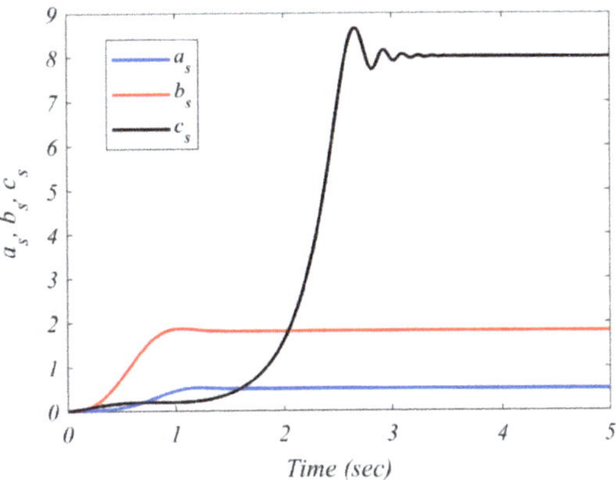

Figure 9. Estimation of uncertain slave parameters corresponding to master trajectories.

As can be seen from Figure 9, that the uncertain slave parameters, a_s and b_s, are rapidly estimated corresponding to the master parameter a_m and b_m, respectively (in less than 1.5 s), which is an exact match to the time duration of the synchronization errors to reach zero. On the other hand, the time duration was about 3.5 s for estimating the third uncertain slave parameter c_s corresponding to the master parameter c_m, where it does not affect the synchronization process. This is because the dynamical behavior of the nonlinear systems is not affected by the same degree of sensitivity for all its parameters as mentioned in [49].

6. Digital Implementation of New Fractional Order Chaotic System

The major goal of the hardware is to test the possibility of implementing fractional order chaotic systems so that they can be employed in real-world applications. Fractional order chaotic systems can be implemented in hardware utilizing a variety of embedded devices, such as microcontrollers, Raspberry Pi, FPGAs, and DSPs as well as implementation by analog electronic circuit as in [29]. In this work, we implemented the new three-dimensional system in Equation (5) by using a microcontroller (Arduino Due) based on a discrete method as in [50].

The Arduino Due is a digital board with Atmel SAM3X8E and ARM Cortex-M3 CPU. It consists of an ideal structure for performing complex arithmetic operations. Briefly, it has the following characteristics: 32-bit ARM core microcontroller; 84 MHz clock; 54 digital input/output pins; 12 analog inputs; USB OTG capable connection; 4 UARTs; 2 DAC (digital to analog); power jack 2 TWI; SPI header; and JTAG header. The primary reasons to use this microcontroller are: 12 bit resolution for its two peripherals, DAC0 and DAC1, and its attractive cost advantage compared with the FPGAs or DSPs boards. An Arduino specified programing language similar to C++ language is used for programing the ARM microcontroller by the Arduino IDE through the native USB port (serial port) [51]. The hardware implementation of the new fractional order chaotic system is shown in Figure 10.

The system parameters for the experimental test are $a = 0.5$, $b = 1.8$, $c = 8$, and $q = 0.99$, with initial conditions ($x(0)$, $y(0)$, and $z(0)$) = (1, 1, 1). The ADC0 and DAC1 are used to give the system in Equation (5) phase portraits of chaotic attractors by analog oscilloscope as displayed in Figure 11. In fact, because the microcontroller's digital to analog converters (DAC0 and DAC1) operate between 0.5 V and 2.7 V, the amplitudes of the simulation results by MATLAB and the experimental results by the Arduino Due will differ for the system's (Equation (5)) state variables (system trajectories). It would be necessary to install an external operational amplification stage in order to achieve the same amplitude values as the computed numerical simulations. Based on the approach that was used for implementing the system (Equation (5)) by the Arduino Due board, there is a level of error of about ±1.56% compared with the simulation results obtained by MATLAB.

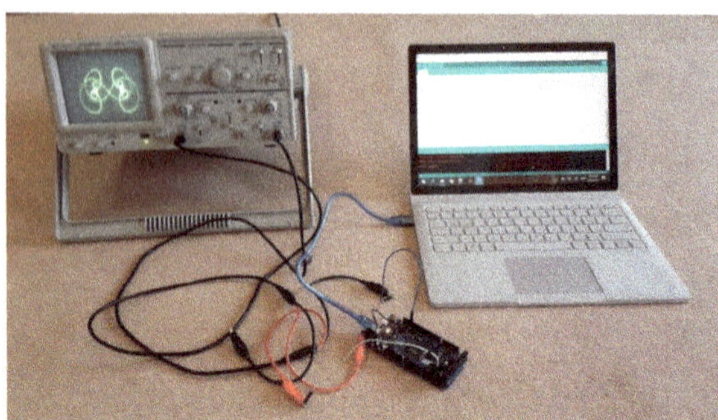

Figure 10. Hardware implementation of the system in Equation (5) based on Arduino Due board.

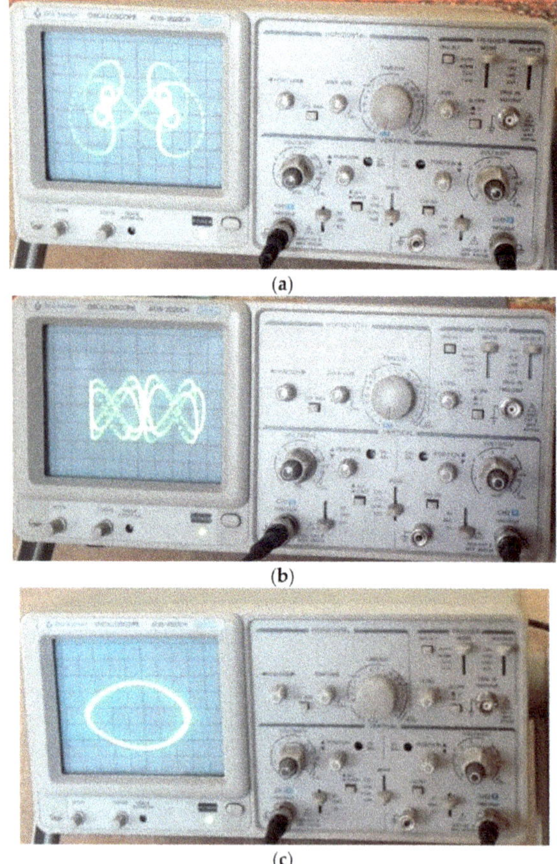

(**a**)

(**b**)

(**c**)

Figure 11. The experimental results of the proposed system phase portraits: (**a**) x-y chaotic attractors; (**b**) y-z chaotic attractors; (**c**) x-z chaotic attractors.

7. Conclusions

A new three-dimensional nonlinear autonomous system with fractional order and chaos exhibition was suggested. The nonlinear dynamical behaviors of this system were analytically and numerically investigated, where these dynamics are the equilibrium points, chaotic attractors, bifurcation diagrams, and Lyapunov exponents. Because the system has no equilibrium, the observed dynamics showed that the system can excite hidden chaotic attractors and display extremely complex dynamics. Afterwards, an adaptive synchronization strategy was formulated. This synchronization approach was set up between two identical new fractional order chaotic systems: one serving as the master and the other as the slave. The adaptive control principles responsible for synchronization verification were derived. Furthermore, in order to estimate the unknown slave parameters, the update laws were determined. Finally, to show the feasibility of using the proposed system in real-world applications, the system was implemented with Arduino Due boards. The obtained numerical results by MATLAB simulation are consistent with the experimental results of the hardware implementation, which show the feasibility of the system to be used in real-world applications in various fields. The main advantage of implementation of this system with Arduino Due boards is the low cost of implementation compared with alternative devices such as FPGAs and DSPs, which require large costs.

Author Contributions: Conceptualization, Z.-A.S.A.R. and B.H.J.; methodology, Z.-A.S.A.R.; software, Z.-A.S.A.R. and B.H.J.; validation, Z.-A.S.A.R. and B.H.J.; formal analysis, Z.-A.S.A.R., B.H.J., and Y.I.A.A.-Y.; investigation, Z.-A.S.A.R. and B.H.J..; resources, Z.-A.S.A.R., B.H.J., Y.I.A.A.-Y., R.A.A.-A. and B.N.A.; data curation, Z.-A.S.A.R. and B.H.J.; writing—original draft preparation, Z.-A.S.A.R. and B.H.J.; writing—review and editing, Z.-A.S.A.R., B.H.J., Y.I.A.A.-Y. and B.N.A.; visualization, Z.-A.S.A.R., B.H.J., Y.I.A.A.-Y. and B.N.A.; supervision, B.H.J. and R.A.A.-A. All authors have read and agreed to the published version of the manuscript.

Funding: This research received no external funding.

Institutional Review Board Statement: Not applicable.

Informed Consent Statement: Not applicable.

Conflicts of Interest: The authors declare no conflict of interest.

References

1. Sordi, A. Chua's oscillator: An introductory approach to chaos theory. *Rev. Bras. Ensino Física* **2021**, *43*. [CrossRef]
2. Vaidyanathan, S.; Volos, C. *Advances and Applications in Chaotic Systems*; Springer: Berlin/Heidelberg, Germany, 2016; Volume 636.
3. Asiain, E.; Garrido, R. Anti-Chaos control of a servo system using nonlinear model reference adaptive control. *Chaos Solitons Fractals* **2021**, *143*, 110581. [CrossRef]
4. Cuellar, G.H.; Jiménez-López, E.; Campos-Cantón, E.; Pisarchik, A. An approach to generate deterministic Brownian motion. *Commun. Nonlinear Sci. Numer. Simul.* **2014**, *19*, 2740–2746. [CrossRef]
5. Jiang, H.; Liu, Y.; Wei, Z.; Zhang, L. Hidden chaotic attractors in a class of two-dimensional maps. *Nonlinear Dyn.* **2016**, *85*, 2719–2727. [CrossRef]
6. Jasim, B.H.; Hassan, K.H.; Omran, K.M. A new 4-D hyperchaotic hidden attractor system: Its dynamics, coexisting attractors, synchronization and microcontroller implementation. *Int. J. Electr. Comput. Eng.* **2021**, *11*, 2068–2078. [CrossRef]
7. Volos, C.K.; Jafari, S.; Kengne, J.; Munoz-Pacheco, J.M.; Rajagopal, K. Nonlinear dynamics and entropy of complex systems with hidden and self-excited attractors. *Entropy* **2019**, *21*, 370. [CrossRef]
8. Azar, A.T.; Serrano, F.E. Stabilization of Port Hamiltonian Chaotic Systems with Hidden Attractors by Adaptive Terminal Sliding Mode Control. *Entropy* **2020**, *22*, 122. [CrossRef]
9. Leonov, G.; Kuznetsov, N.V.; Mokaev, T. Homoclinic orbits, and self-excited and hidden attractors in a Lorenz-like system describing convective fluid motion. *Eur. Phys. J. Spéc. Top.* **2015**, *224*, 1421–1458. [CrossRef]
10. Leonov, G.; Kuznetsov, N.V.; Kiseleva, M.; Solovyeva, E.P.; Zaretskiy, A.M. Hidden oscillations in mathematical model of drilling system actuated by induction motor with a wound rotor. *Nonlinear Dyn.* **2014**, *77*, 277–288. [CrossRef]
11. Sambas, A.; Vaidyanathan, S.; Bonny, T.; Zhang, S.; Sukono; Hidayat, Y.; Gundara, G.; Mamat, M. Mathematical Model and FPGA Realization of a Multi-Stable Chaotic Dynamical System with a Closed Butterfly-Like Curve of Equilibrium Points. *Appl. Sci.* **2021**, *11*, 788. [CrossRef]
12. Andrievsky, B.; Kuznetsov, N.; Leonov, G.; Pogromsky, A. Hidden oscillations in aircraft flight control system with input saturation. *IFAC Proc. Vol.* **2013**, *46*, 75–79. [CrossRef]
13. Pham, V.T.; Jafari, S.; Vaidyanathan, S.; Volos, C.; Wang, X. A novel memristive neural network with hidden attractors and its circuitry implementation. *Sci. China Ser. E Technol. Sci.* **2015**, *59*, 358–363. [CrossRef]
14. Kiani-B, A.; Fallahi, K.; Pariz, N.; Leung, H. A chaotic secure communication scheme using fractional chaotic systems based on an extended fractional Kalman filter. *Commun. Nonlinear Sci. Numer. Simul.* **2009**, *14*, 863–879. [CrossRef]

15. Kuznetsov, N.V.; Leonov, G.A.; Yuldashev, M.V.; Yuldashev, R.V. Hidden attractors in dynamical models of phase-locked loop circuits: Limitations of simulation in MATLAB and SPICE. *Commun. Nonlinear Sci. Numer. Simul.* **2017**, *51*, 39–49. [CrossRef]
16. Vlachas, P.R.; Byeon, W.; Wan, Z.Y.; Sapsis, T.P.; Koumoutsakos, P. Data-driven forecasting of high-dimensional chaotic systems with long short-term memory networks. *Proc. R. Soc. A Math. Phys. Eng. Sci.* **2018**, *474*, 20170844. [CrossRef]
17. Velamore, A.A.; Hegde, A.; Khan, A.A.; Deb, S. Dual cascaded Fractional-order Chaotic Synchronization for Secure Communication with Analog Circuit Realisation. In Proceedings of the 2021 IEEE Second International Conference on Control, Measurement and Instrumentation (CMI), Kolkata, India, 8–10 January 2021; IEEE: New York, NY, USA, 2021; pp. 30–35.
18. Pham, V.-T.; Vaidyanathan, S.; Volos, C.K.; Jafari, S. Hidden attractors in a chaotic system with an exponential nonlinear term. *Eur. Phys. J. Spéc. Top.* **2015**, *224*, 1507–1517. [CrossRef]
19. Tahir, F.R.; Jafari, S.; Pham, V.-T.; Volos, C.; Wang, X. A Novel No-Equilibrium Chaotic System with Multiwing Butterfly Attractors. *Int. J. Bifurc. Chaos* **2015**, *25*. [CrossRef]
20. Pham, V.-T.; Vaidyanathan, S.; Volos, C.; Kapitaniak, T. *Nonlinear Dynamical Systems with Self-Excited and Hidden Attractors*; Springer: Berlin/Heidelberg, Germany, 2018; Volume 133.
21. Bayani, A.; Rajagopal, K.; Khalaf, A.J.M.; Jafari, S.; Leutcho, G.D.; Kengne, J. Dynamical analysis of a new multistable chaotic system with hidden attractor: Antimonotonicity, coexisting multiple attractors, and offset boosting. *Phys. Lett. A* **2019**, *383*, 1450–1456. [CrossRef]
22. Rahman, Z.-A.S.A.; Jassim, B.H.; Al-Yasir, Y.I.A. New Fractional Order Chaotic System: Analysis, Synchronization, and it's Application. *Iraqi J. Electr. Electron. Eng.* **2021**, *17*. [CrossRef]
23. Escalante-González, R.J.; Campos, E. Multistable systems with hidden and self-excited scroll attractors generated via piecewise linear systems. *Complex. Dyn. Control Appl. Nonlinear Syst. Multistability* **2020**, *2020*. [CrossRef]
24. Ma, C. A Novel Computational Technique for Impulsive Fractional Differential Equations. *Symmetry* **2019**, *11*, 216. [CrossRef]
25. Sun, H.; Zhang, Y.; Baleanu, D.; Chen, W.; Chen, Y. A new collection of real world applications of fractional calculus in science and engineering. *Commun. Nonlinear Sci. Numer. Simul.* **2018**, *64*, 213–231. [CrossRef]
26. Boubaker, O.; Jafari, S. *Recent Advances in Chaotic Systems and Synchronization: From Theory to Real World Applications*; Academic Press: Amsterdam, The Netherlands, 2018.
27. Abdelaty, A.M.; Azar, A.T.; Vaidyanathan, S.; Ouannas, A.; Radwan, A. Applications of Continuous-time Fractional Order Chaotic Systems. *Math. Tech. Fract. Order Syst.* **2018**, 409–449. [CrossRef]
28. Soleimanizadeh, A.; Nekoui, M.A. Optimal type-2 fuzzy synchronization of two different fractional-order chaotic systems with variable orders with an application to secure communication. *Soft Comput.* **2021**, *25*, 6415–6426. [CrossRef]
29. Echenausía-Monroy, J.; Gilardi-Velázquez, H.; Jaimes-Reátegui, R.; Aboites, V.; Huerta-Cuellar, G. A physical interpretation of fractional-order-derivatives in a jerk system: Electronic approach. *Commun. Nonlinear Sci. Numer. Simul.* **2020**, *90*, 105413. [CrossRef]
30. Lai, Q.; Wan, Z.; Kuate, P.D.K.; Fotsin, H. Coexisting attractors, circuit implementation and synchronization control of a new chaotic system evolved from the simplest memristor chaotic circuit. *Commun. Nonlinear Sci. Numer. Simul.* **2020**, *89*, 105341. [CrossRef]
31. Rahman, Z.-A.S.A.; Al-Kashoash, H.A.A.; Ramadhan, S.M.; Al-Yasir, Y.I.A. Adaptive Control Synchronization of a Novel Memristive Chaotic System for Secure Communication Applications. *Inventions* **2019**, *4*, 30. [CrossRef]
32. Jasim, B.H.; Mjily, A.H.; AL-Aaragee, A.M.J. A novel 4 dimensional hyperchaotic system with its control, synchronization and implementation. *Int. J. Electr. Comput. Eng.* **2021**, *11*, 2974–2985.
33. Agrawal, S.; Srivastava, M.; Das, S. Synchronization of fractional order chaotic systems using active control method. *Chaos Solitons Fractals* **2012**, *45*, 737–752. [CrossRef]
34. Tavazoei, M.S.; Haeri, M. Synchronization of chaotic fractional-order systems via active sliding mode controller. *Phys. A Stat. Mech. Appl.* **2008**, *387*, 57–70. [CrossRef]
35. Kumar, S.; Matouk, A.E.; Chaudhary, H.; Kant, S. Control and synchronization of fractional-order chaotic satellite systems using feedback and adaptive control techniques. *Int. J. Adapt. Control. Signal Process.* **2021**, *35*, 484–497. [CrossRef]
36. Liu, L.; Du, C.; Zhang, X.; Li, J.; Shi, S. Adaptive Synchronization Strategy between Two Autonomous Dissipative Chaotic Systems Using Fractional-Order Mittag–Leffler Stability. *Entropy* **2019**, *21*, 383. [CrossRef]
37. Megherbi, O.; Hamiche, H.; Djennoune, S.; Bettayeb, M. A new contribution for the impulsive synchronization of fractional-order discrete-time chaotic systems. *Nonlinear Dyn.* **2017**, *90*, 1519–1533. [CrossRef]
38. Ortigueira, M.D.; Machado, J.T.; Trujillo, J.J. Fractional derivatives and periodic functions. *Int. J. Dyn. Control.* **2017**, *5*, 72–78. [CrossRef]
39. Huseynov, I.T.; Ahmadova, A.; Fernandez, A.; Mahmudov, N.I. Explicit analytical solutions of incommensurate fractional differential equation systems. *Appl. Math. Comput.* **2021**, *390*, 125590. [CrossRef]
40. Wang, C.; Ding, Q. A New Two-Dimensional Map with Hidden Attractors. *Entropy* **2018**, *20*, 322. [CrossRef] [PubMed]
41. Jiang, Y.; Zhang, B. Comparative Study of Riemann–Liouville and Caputo Derivative Definitions in Time-Domain Analysis of Fractional-Order Capacitor. *IEEE Trans. Circuits Syst. II Express Briefs* **2020**, *67*, 2184–2188. [CrossRef]
42. Zheng, Z.; Zhao, W.; Dai, H. A new definition of fractional derivative. *Int. J. Nonlinear Mech.* **2019**, *108*, 1–6. [CrossRef]
43. He, S.; Sun, K.; Banerjee, S. Dynamical properties and complexity in fractional-order diffusionless Lorenz system. *Eur. Phys. J. Plus* **2016**, *131*, 1–12. [CrossRef]
44. Garrappa, R. Numerical Solution of Fractional Differential Equations: A Survey and a Software Tutorial. *Mathematics* **2018**, *6*, 16. [CrossRef]

45. Qi, G.; Xu, L.; Yang, X. Energy mechanism analysis for chaotic dynamics of gyrostat system and simulation of displacement orbit using COMSOL. *Appl. Math. Model.* **2021**, *92*, 333–348. [CrossRef]
46. Stankevich, N.V.; Kuznetsov, A.P.; Seleznev, E.P. Chaos and hyperchaos arising from the destruction of multifrequency tori. *Chaos Solitons Fractals* **2021**, *147*, 110998. [CrossRef]
47. Xu, Y.; Wang, H.; Li, Y.; Pei, B. Image encryption based on synchronization of fractional chaotic systems. *Commun. Nonlinear Sci. Numer. Simul.* **2014**, *19*, 3735–3744. [CrossRef]
48. Fan, X.; Wang, Z. A Fuzzy Lyapunov Function Method to Stability Analysis of Fractional Order T-S Fuzzy Systems. *IEEE Trans. Fuzzy Syst.* **2021**, *PP*, 1. [CrossRef]
49. Nowak, R.D. Nonlinear system identification. *Circuits Syst. Signal Process* **2002**, *21*, 109–122. [CrossRef]
50. Sánchez-López, C. An experimental synthesis methodology of fractional-order chaotic attractors. *Nonlinear Dyn.* **2020**, *100*, 3907–3923. [CrossRef]
51. Due, A. Core ARM. Arduino Due. *Retrieved* **2017**, *9*, 2019.

Article

A Portable Low-Cost Ultrasound Measurement Device for Concrete Monitoring

Daniel Fontoura Barroso *, **Niklas Epple** and **Ernst Niederleithinger**

Bundesanstalt für Materialforschung und -Prüfung (BAM), Unter den Eichen 87, 12205 Berlin, Germany; Niklas.Epple@bam.de (N.E.); Ernst.Niederleithinger@bam.de (E.N.)
* Correspondence: daniel.barroso@bam.de; Tel.: +49-030-8104-4398

Abstract: This paper describes a new ultrasonic measuring device called "W-Box". It was developed based on the requirements of the DFG Forschergruppe (research unit) CoDA for a portable device for monitoring of concrete specimens, models and actual structures using embedded ultrasonic transducers as well as temperature and humidity sensors. The W-Box can send ultrasonic pulses with a variable frequency of 50–100 kHz to one selectable transducer and records signals from up to 75 multiplexed channels with a sample rate of 1 MHz and a resolution of 14 bits. In addition, it measures temperature and humidity with high accuracy, adjustable amplification, restarts automatically after a power failure and can be fully controlled remotely. The measured data are automatically stored locally on-site data quality checks and transferred to remote servers. The comparison of the W-Box with a laboratory setup using commercial devices proves that it is equally reliable and precise, at much lower cost. The W-Box also shows that their measurement capacities, with the used embedded ultrasonic transducers, can reach above 6 m in concrete.

Keywords: low-cost; ultrasound; IoT; non-destructive testing; coda wave interferometry

Citation: Fontoura Barroso, D.; Epple, N.; Niederleithinger, E. A Portable Low-Cost Ultrasound Measurement Device for Concrete Monitoring. *Inventions* **2021**, *6*, 36. https://doi.org/10.3390/inventions6020036

Academic Editors: Francisco Manzano Agugliaro and Esther Salmerón-Manzano

Received: 7 April 2021
Accepted: 10 May 2021
Published: 20 May 2021

1. Introduction

Monitoring and non-destructive testing (NDT) has become a major topic in civil engineering (CE) in recent decades, as safety and environmental issues have shown the need for methods to ensure the structural integrity and increase the lifespan of constructions such as bridges. One of the most researched and widely spread methods of NDT is measuring with ultrasound. With the increasing demand for ultrasound measurements—especially for long-term monitoring—the need for a cost efficient, easy to build, and reliable measurement system has been identified by researchers at the NDT-CE department of the German Federal Institute for Materials Research and Testing (BAM).

Ultrasound transmission experiments for concrete are considered as a standard procedure for quality assessment. Time of flight (TOF) measurements with compressional waves are applied in a standardized manner to estimate concrete strength using ultrasonic velocity as a proxy [1]. The method is also used to follow the hardening of fresh concrete [2]. In research, ultrasound is used to monitor changes in concrete in controlled experiments by following changes certain features such as velocity and/or amplitudes, e.g., to assess fatigue strength [3].

Since around 2000, several researchers have started to extend the capabilities of ultrasonic monitoring either by using more sensitive data evaluation techniques. This includes but is not limited to coda wave interferometry (CWI) or nonlinear acoustic features. Several laboratory experiments have been published in recent years showing that ultrasonic monitoring combined with these newly developed sensitive feature extraction techniques are very sensitive to stress, temperature, moisture, cracking and any material changes in general, even if they are minute or local. The potential of these approaches and progress of the early years has been compiled in several publications (see [4–6]). The background of this research is to provide additional means for structural health monitoring of degrading concrete structures such as transportation infrastructure or in nuclear power generation and waste storage.

Recently the results of several experiments have been published which show the potential of CWI and other techniques for practical implementation (see [7–10]). In these experiments either commercial transducers mounted on the surface [7] or special embedded transducers [11] have been applied. For signal generation and data acquisition customized setups of commercial instruments have been applied so far. Most of them have limitations in terms of cost, ruggedness and availability. In the frame of the DFG FOR 2825 CoDA research unit, a collaboration of several German universities and BAM working on the improvement of CWI for concrete monitoring, the need arose for a compact, robust and affordable measurement device which can be built at university labs with limited resources in significant numbers. The device should not be limited to the purpose of the research unit, but also be useful in traditional ultrasonic transmission experiments. It will be made available in public the domain.

In cooperation with students from TU Berlin, a first prototype of a 5-channel ultrasound measurement system based on a Raspberry Pi was developed in 2015 [12]. This prototype, called the W-Box due to its initial design, was further developed and improved over the years and is now being used by the research unit "CoDA" funded by the German Research Foundation (DFG) to investigate the changes in ultrasonic signals in reinforced concrete structures in order to detect changes and damage. Within the project, the system was extended to connect to up to 75 ultrasonic transducers, options for temperature and humidity measurements were added, automatic data upload and remote control were introduced, and the system was made more resistant to environmental effects such as rain or high humidity. In this paper, we will explain the design of the W-Box by describing the requirements and the individual components of the system. Detailed descriptions of each electronic board are presented, followed by a comparison between the measurement results acquired with the W-Box and those acquired with a commercially available measurement system, such as the system used in [11,13], which has been used in the past at BAM.

2. Materials and Methods

In the following, we will explain the design of the W-Box, the required electronic components, and their characteristics. Firstly, in Sections 2.1 and 2.2, the general requirements posed to the system by the ultrasonic measurement task and the system design are explained. This is followed by a detailed explanation of the shared Raspberry Pi pins connection necessary for communication between the individual electronic components (Section 2.3). The short introduction to the Raspberry Pi and the operating system (Section 2.4) is followed by a detailed description of the individual boards of the measurement system (Sections 2.5–2.9).

2.1. System Requirements

The W-Box originated from the need for portable equipment with which to carry out ultrasound monitoring. Therefore, it needed to be small, light, easy to operate, simple to repair, and inexpensive. This would allow it to be used in variable locations which are often remote or hard to access. As several sensors are usually required for monitoring, the W-Box should be capable of connecting to at least 20 ultrasonic transducers, all operating as transmitters and receivers, adjustable signal amplification, switching between them using multiplexers and sending variable pulses of up to 300 Vpp. It also must be capable of restarting measurements after power outage, of saving data onto an external server, and being protected against high humidity (e.g., for operation in climate chambers). As ultrasonic wave propagation is influenced by temperature and moisture content in addition to the main objectives (damage, deterioration), additional capacities for connecting of temperature and humidity sensors had to be added to allow compensation. The main features and parameters are listed in Table 1.

Table 1. W-Box main project features.

Ultrasound channels	up to 75
Pulse frequency	from 20 kHz up to 100 kHz
Pulse voltage	up to 300 Vpp
Analog Filter	low-pass at 150 kHz
Amplification	from 1× up to 150×
Humidity	1 channel
Temperature	up to 3 channels
Remote control	yes

2.2. System Design

The system is designed around a Raspberry Pi 3 B+ and runs with an Ubuntu operating system. The W-Box is capable of providing a I2C protocol for communication between all micro-controllers, of connecting to the internet (because of embedded Wi-Fi or an RJ45 internet port), of connecting to an external server, of automatically restarting after power outage, and of being easily and fully remote-controlled, if required.

Figure 1 shows the latest version of the W-Box. It has 20 connectors for ultrasonic transducers (BNC female). The temperature and humidity board are in the final stage of development. It is capable of measuring with 3 NTC temperature sensors and 1 humidity sensor. The black case is a standard case and the frontal pink board is 3D printed. All circuits have the same Raspberry Pi pins between them in order to supply power and to communicate as described in Figure 2. Its communication functions are explained in Figure 3.

In Figure 3, a schematic of the communication between all systems is shown. First, the Raspberry Pi sends the desired values for measurement to all boards, except the pulser board, via I2C. The power board provides the selected transmitter output voltage. The multiplexer prepares the chosen input/output combination of the ultrasonic sensors. The temp/humidity board sends the acquired values back to the Raspberry Pi via I2C. After that, the whole system is prepared for the pulse. The ADC board sends the digital pulse to the pulser board (orange arrow in Figure 3), which forwards it to the selected transmitting transducer and records the input signal (green arrow in Figure 3). After the ultrasonic signal from the currently selected receiving transducer is recorded on the ADC board, it is digitally sent to the Raspberry Pi via I2C. This measurement sequence takes around four seconds, depending on the number of samples recorded.

Figure 1. The W-Box. The 3D-printed pink side panel holds 20 BNC transducer connectors.

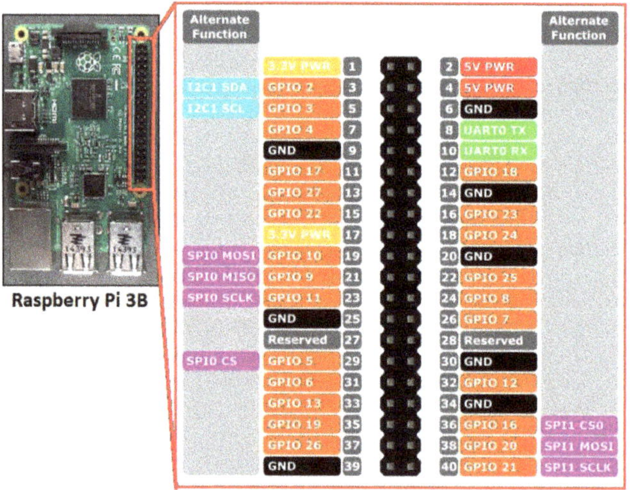

Figure 2. Schematic zoom of the Raspberry Pi 3 B+ pins. Adapted from [14].

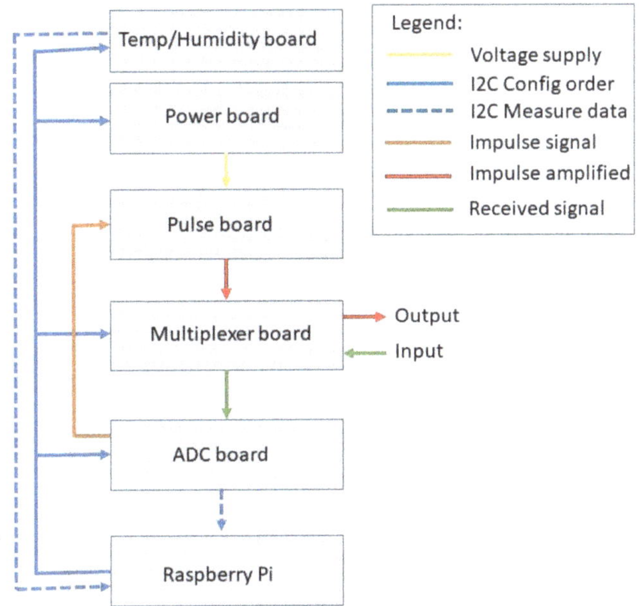

Figure 3. W-Box communication schematic. The gray boxes represent each circuit of the W-Box and the colored arrows the communication between them.

2.3. Protocol and Electric Communication Design

There are 3 electric connections between the boards. The first connection's only function is to deliver the required voltage supply from the power board to the pulser board (Figure 4 number 2). The second connection delivers the pulse from the pulser board, via the multiplexer, to the sending transducer (Figure 5 number 1). The third connection opens the circuit for the measurement data; this travels from the receiving transducer through the multiplexer to the ADC board (Figure 4 number 1).

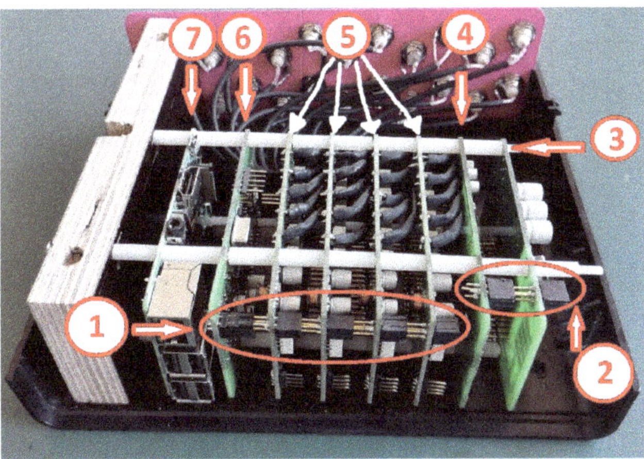

Figure 4. Interior view of the W-Box, after removing the top and rear cover. 1 and 2: Electronic connections between the circuits. 3: Power board. 4: Pulser board. 5: Four multiplexer boards. 6: ADC (Analog-to-digital converter) board. 7: Raspberry Pi.

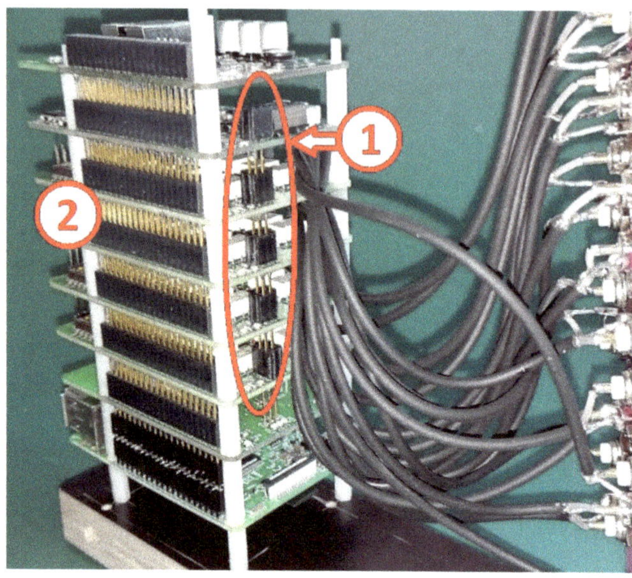

Figure 5. Rear photo of the W-Box electronics without the protective case. Right: rear side of BNC connectors. 1: Electrical connections between the boards. 2: Pin connections trough all boards.

In order to clarify how the boards communicate and how they are connected with each other, all data/electric connections between the boards and the Raspberry Pi will be explained in the following. As shown in Figure 2, it provides the 5 V and 3.3 V needed for all circuits. GPIO 2 and 3 support the I2C communication protocol between each main microcontroller and the Raspberry Pi, allowing all the predetermined functions to be triggered and the measured data to be received after each measurement. This is also shown later in Figure 12 in the final section as a continuous blue line.

GPIO 16, 20 and 21 are used to trigger the pulse to the pulser board. This digital trigger is controlled by the microcontroller TM4C123GH6PM, which sends two digital pulses with a predetermined length defining the center frequency. This will be explained in Section 2.8.

The Raspberry Pi extension pins are shown in Figure 5, number 2. There are also three electrical connections out of the pins. They are shown Figure 4, numbers 1 and 2, and also in Figure 5, number 1.

Figure 4 shows an internal front view of the W-Box. Number 1 shows a section containing plates indicating the connection between the ultrasound input and the ADC board, via the multiplexers. Number 2 highlights the connection between the power board and the pulser board. Numbers 3, 4, 5, 6, and 7, indicate, respectively, the power board, the pulser board, the 4 multiplexer boards, the ADC board, and the Raspberry Pi.

Figure 5 shows an internal rear view of the W-Box. Number 1 indicates the connection between the pulser board and all 4 multiplexer boards. Number 2 indicates the Raspberry Pi pins explained in Figure 2. Although UART0 TX, UART0 TX, GPIO 9, GPIO 10 and GPIO 11 are connected into the ADC board, they are not yet in use. However, these circuit connections are prepared for possible future protocol communication enhancements.

2.4. *Software and Ubuntu OS*

There are 3 main functions for the OS and the control software developed in Python. The first is to provide a graphic interface for the user (GUI), the second is to save the measured data in its own memory or on the internet, and the third is to send commands and receive data via the I2C protocol. As soon as the W-Box is connected to the power source, the operating system (Ubuntu) is booted and the ultrasound program automatically starts a measurement with the previously prepared configuration. After an eventual power failure, the W-Box restarts automatically and continues to acquire and store/send data.

Figure 6 shows the GUI after a single test between two embedded ultrasonic transducers (S0807, ACS International). The amplitude of the signal (from 7000 to 8500) is within the 16,384 possible values of a 14-bit ADC. The user can customize how the data are sent to an external server. He can as well access the internal memory via VNC and copy the saved data using Ubuntu automatic servers or native Linux VNC controllers. A "configuration.ini" file defines the parameters for the repeated and the automatically started measurements. The data are stored as integer values in a single file for every measurement.

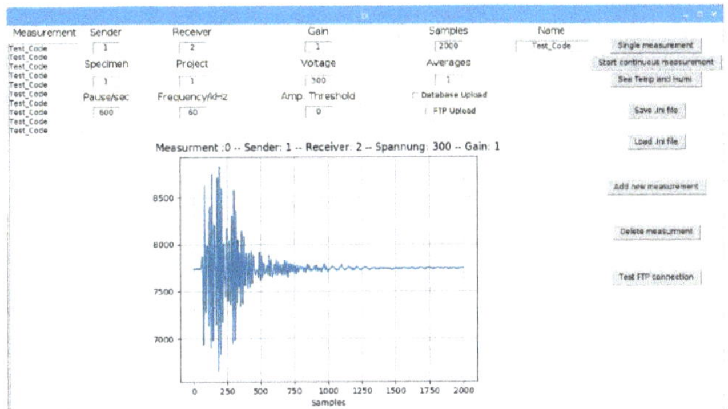

Figure 6. User interface of the measurement program installed on the Raspberry Pi. The settings for all measurements can be adapted individually. Data quality can be checked using the plot of an ultrasound signal.

2.5. Multiplexer Board

The function of the multiplexer is to define the channels for transmitting and receiving ultrasonic signals to and from the transducers. The board is directly controlled by the Raspberry Pi via I2C and switches between the output and input connectors before the transmitter impulse signal is sent.

In order to allow for additional multiplexer boards as well as easier testing, a DIP switch (see Figure 7, number 7) is integrated on each multiplexer board. In this way, it is possible to manually change the address of each board between 0000 and 1110, allowing the use of up to 15 boards. When set to 1111, it automatically initiates the test mode. In test mode, all input and output channels of the multiplexer board are tested. The functionality of the W-Box can be easily checked with a multimeter. LEDs (Number 3) are integrated at the top of each relay (Number 4) to show if they are closed. Additional two LEDs at the top left of the board (Number 1) confirm that the 5 V and 3.3 V circuit currents are flowing through the board. They are protected against short circuit with a 2 A and 1 A fuse; (Number 1, green and blue SMD components). Number 2 in Figure 7 shows the connection to the ADC. Number 5 shows the connection to the multiplexer. Number 6 is the programmable microcontroller (MSP430G2303 from Texas Instruments) which receives I2C commands from the Raspberry Pi and digitally controls all multiplexer board functions. Number 8 shows the five connections to the BNC connectors.

Figure 7. Multiplexer board, front view. 1: Fuse: 2. Connection to the pulser board. 3: LED. 4: Relay. 5: Connection to the ADC board. 6: Microcontroller. 7: Dip switch. 8: Output cable connection to the BNC connectors.

2.6. ADC Board

The ADC board is dedicated to acquire data and send a trigger pulse to the pulser board. The measurements begin when the trigger is sent, which goes through the GPIO 20 and GPIO 21, with positive and negative trigger signaling, respectively (see Figure 12).

On the ADC board, there are 2 main components (see Figure 8, numbers 1 and 2); both have multiple connections in the middle. The first (TM4C123GH6PM from Texas Instruments), number 1, is a microcontroller programmed to receive and send I2C commands and to send the digital pulse to the pulser board. It also controls the second main component, number 2, which is an analog-to-digital converter (AD9241ASZ from Analog Devices Inc.) responsible for receiving the analog data and sending it back digitally to number 1. Number 4 shows the analog ultrasound signal received from the multiplexer. Before this signal goes into the ADC (Number 2), it is amplified and filtered in an analog filtering circuit (Number 5), as explained in the filtering and amplification paragraph.

Considering that the AD9241ASZ has 14 bits and is adjusted to measure between ±5 V, the resolution of the system without amplification is 610 μV.

Figure 8. ADC board, front view. 1: Microcontroller. 2: Analog-to-digital converter. 3: 1Mhz crystal oscillator. 4: Connection to the multiplexer board. 5: Analog filter electronics position.

Filtering and Amplification

The received signal is electronically amplified for each individual measurement by a factor of up to 150. The amplification gain is controlled by a variable resistor, which is controlled by the ADC board microprocessor.

To prevent aliasing effects and to reduce noise, the received analog signal is filtered after amplification. The analog filter is located between the analog input to the ADC (Figure 8, number 2) and the output from the multiplexer (Figure 7, number 4). The filter and amplifier electronics are positioned on the ADC board, as shown in Figure 8, number 5. The filter used is a low-pass 4th order Bessel-Filter. This filter was chosen because of the small amount of overshoot compared to other filters, meaning that the signal will preserve the wave shape of filtered signals in the passband.

Figure 9 shows the calculated frequency response of the designed Bessel-Filter. As the system is supposed to work with piezoceramic ultrasonic sensors with a 60 kHz center frequency, the filter design results are in line with expectations. The input calculated specifications are: +Vs = 5 V, Vs = 0, GBW (gain–bandwidth product). Filter specifications are: Gain: 0 dB. Passband: −3 dB at 150 kHz, Stopband: −40 dB at 800 kHz. Component tolerances are: capacitor: 1 %; resistor: 0.5 %; inductor: 5 %; op amp GBW: 20 %.

2.7. Power Board

The power board is responsible for providing the desired voltage for the pulser board. It receives information from the Raspberry Pi via an I2C protocol, produces the chosen voltage, and is deactivated after the measurements.

In Figure 10, number 1, there are 2 LEDs and fuses to indicate the normal energy supply and to protect the 5 V and 3.3 V circuits. The blue fuse is for protection of up to 1 A for the 3.3V circuit and the green is for 2 A for the 5 V circuit. The other LEDs in Figure 10 (Numbers 2 and 3) serve to indicate the functionality in other circuits. Number 4 is the electronic connection to the pulser board. Number 5 is the transformer for the voltage capacitor charger controlled by the controller (Model LT3751 from Analog Devices), which is positioned in number 6. The main controller is the same as for the multiplexer board (MSP430G2303, number 7).

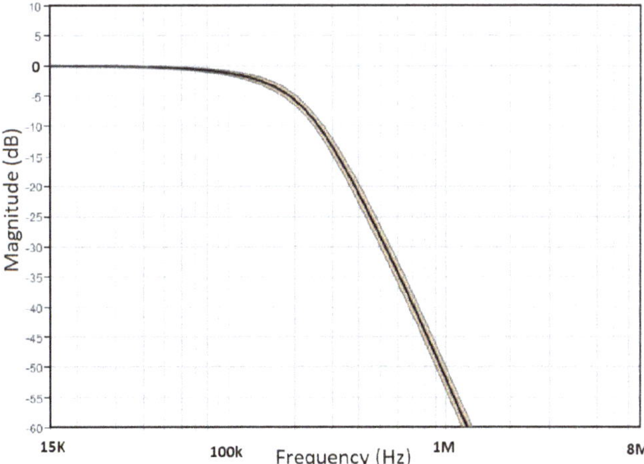

Figure 9. ADC board 4th order Bessel−Filter, calculation of frequency response.

Figure 10. Power board, top view. 1, 2 and 3: LED's indicators. 4: Connection to the pulser board. 5: DC/DC converter. 6 and 7: Microcontrollers.

2.8. Pulser Board

The pulser board receives the user selected voltage from the power board and sends the pulse to the multiplexer board via the connectors indicated in Figure 11, number 2. The pulse frequency and triggering are digitally controlled from the ADC board. These commands are sent through the GPIO 20 with positive pulse reference and through the GPIO 21 with negative pulse reference, while the GPIO 16 is the trigger reference (GPIO 16, 20, 21, as seen in Figure 2). Briefly explained, the board is just waiting for a trigger signal to pulse the signal through the multiplexer. This board has no microprocessor due to its purely analogue function to replicate and amplify the received digital signal from the trigger input.

Figure 11, number 1 shows two LEDs which indicate whether energy is going through the 5 V and 3.3 V circuits. These circuits are also protected with the same fuse as the power board and multiplexer. These fuses are positioned at the left side of the LEDs. Number 2 is the connection to the multiplexer board and number 3 is the power voltage supply connection from the power board.

Figure 11. Pulser board, top view. 1: LED's: 2. Connection to the multiplexer board. 3: Connection to the power board.

Pulse Signal

The square wave transmit pulse has been used in [15–17]. It is also widely used by commercial ultrasound measurement equipment. Because of the simplicity, comparability and ease construction of the pulse electronics, the square wave transmit pulse was chosen in this project as well. The format of the sent impulse is detailed in Figure 12, where it is labeled as the "Output pulse" and shown in green.

The length of the positive and negative parts of the output pulse are controlled by the digital signals GPIO 20 and GPIO 21, respectively, from the ADC board. These digital signals are shown in Figure 12 in blue and red, respectively. The analog pulse frequency and amplitude can be modified from 20 kHz up to 100 kHz, and the voltage up to ±300 V.

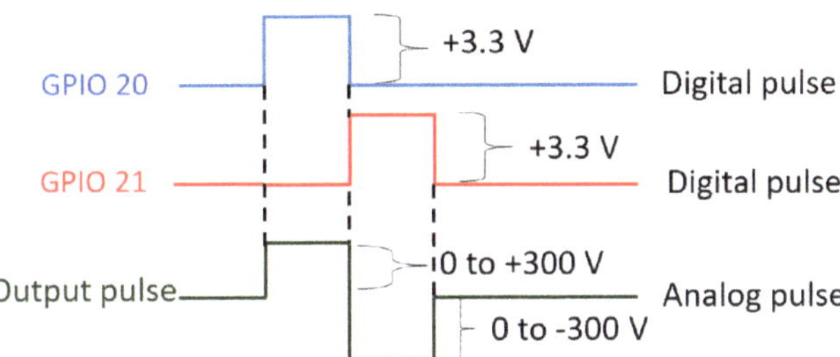

Figure 12. The analog output pulse (green) sent to the transducer is provided by a combination of two digital pulses (blue and red) created by GPIO 20 and GPIO 21.

2.9. Temperature and Humidity Board

The temperature and humidity board is in its last stage of development. It can acquires values data from up top three temperature NTC 10K Ω sensors and one humidity sensor in each measurement cycle.

As can be seen in Figure 13, there are 5 outputs in the temperature board. Numbers 1, 2, and 3 show the connection to the temperature sensors. Number 4 (orange and brown cables) is the reference to the humidity sensor and number 5 (red and black cables) provides a specific voltage to the humidity sensor. Number 6 is the DIP switch in 1111, the I2C address of the board. Number 7 is the controller MSP430G2333.

The chosen temperature sensor (TT0210KC3-T105-1500 from TEWA sensors) was also used in the experiment described in [18]. It is weatherproof (IP68) and thus ready to be embedded into wet concrete due to stranded tinned copper and insulation. It works accurate between −50 °C and +105 °C, and only loses precision up to 0.02 °C per year, whereas a normal NTC sensor loses up to 0.2 °C per year. The humidity sensor chosen for our experiment is the Honeywell HIH-5031 because of its low energy demand, linearity, reliability, and an embedded hydrophobic filter. According to the NTC thermo sensor specifications and its developed electronics, it is possible to calculate its repeatability, within the range of −10 °C and +40 °C, of 0.5 °C for the temperature value. The accuracy has not been determined yet. The specified precision of the humidity sensor is 0.5%, and its accuracy is about 3% of the relative humidity values.

Figure 13. Temperature and humidity board prototype. 1, 2 and 3: Temperature sensors output. 4 and 5: Output connection to the humidity sensor. 6: DIP switch. 7: Microcontroller.

3. Results

In order to prove the functionality and accuracy of the W-Box, we have compared the measurements acquired with this novel system to a state-of-the-art industrial grade measurement system. At BAM, US measurements are typically conducted with a National Instruments DAQmx 6361 ADC, a custom build pulse generator, a Keithley 2700 multiplexer and a Stanford Research Amplifier as described in [13]. In the following, we will compare results from both measurement systems and give an estimation of the measurement quality of the novel W-Box system. For this experiment, we used a concrete prism (40 × 10 × 10 cm) with two embedded ultrasonic sensors (see Figures 14 and 15), the same as those used in [11,13,18]. In the following, the results from the NI system displayed in Figure 16 will be called "classic measurement", or just "classic", when comparing them to the measurements recorded with the W-Box.

In Figure 16, the classic measurement equipment is shown together with the computer and the measurement program: an analog signal amplifier (1), DAQmx (2), and Keithley 2700 multiplexer (3). Even though the DAQmx can measure up to 2 MHz, for this comparison experiment it was adjusted to 1 MHz, the same sampling frequency as the W-Box. It is also important to acknowledge that DAQmx can measure with 16-bit and the W-Box with 14-bit resolution. This results in a precision difference of 2 bits, resulting in 4 times more precision in amplitude for DAQmx measurements. The maximum pulse capacity is 150 Vpp for Classic, and 300 Vpp for the W-Box.

Figure 14. Concrete beam of 40 cm × 10 cm × 10 cm and two embedded ultrasonic sensors (S0807 from ACSYS, as used in [11,13,18,19]).

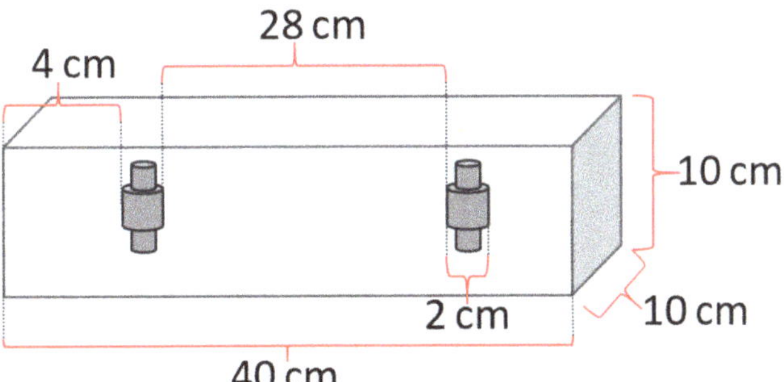

Figure 15. Schematic drawing of the beam of Figure 14. In the center of the image are the two embedded ultrasonic sensors. Both sensors are aligned in the same depth and lateral distances.

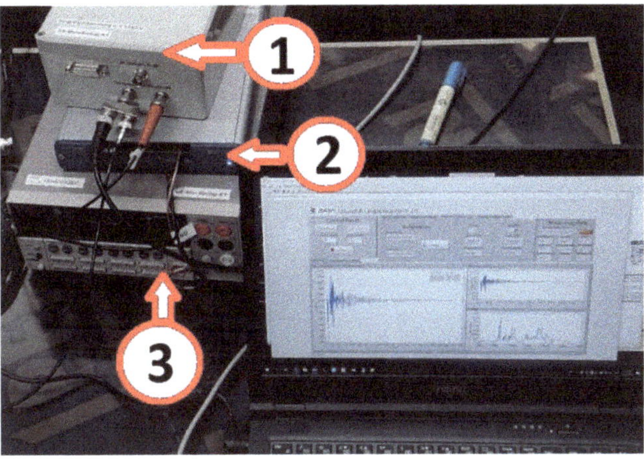

Figure 16. BAM Classic ultrasound measurement system setup. 1: Pulse generator. 2: DAQmx. 3: Keithley Multiplexer. Notebook screen shows the NI LabView control software GUI.

Data acquired using both classic instrumentation and the W-Box are pre-processed by a Python script to remove the offset, suppress crosstalk (crosstalk is expected at the beginning of each measurement because of the electric pulse), to perform digital filtering with a band-pass filter from 10 kHz up to 150 kHz (based on the expected frequency range from our transducers), and to remove the pre-trigger samples (first 50 samples for the W-Box and 100 samples for classic instrumentation). The signals are averaged 24-fold, temporally interleaved between both setups, for a period of approximately one hour at the same temperature and humidity.

For normalization of the graphs in Figure 17, the maximum received voltage values from both systems were acquired, showing the maximal voltage value of 0.270 Vpp for Classic and 0.513 Vpp for the W-Box. To compare the signals, it is necessary to use a regular method of error. According to [20], the root-mean-square error (RMSE) is a widely used equation to compare a measurement with a predicted model or reference measurement. Additionally, according to [21], some researchers recommend the use of the Mean Absolute Error (MAE) instead of RMSE. The method has some interpretability advantages over the RMSE. MAE is fundamentally the average magnitude of the errors and it is also easier to understand. One of MAE's main characteristics is that each error influences in direct proportion to the absolute value of the error, which is not the case for RMSE. Due to this, both methods will be used. Both errors are calculated after the normalization and adjustment of the two graphs in order to facilitate correct error correlation calculation.

$$\text{RMSE} = \sqrt{\frac{\sum_{i=1}^{n} (y_i + x_i)^2}{n}} = 0,040 \quad (1)$$

The square roots of the differences between Classic measurements values and W-Box values for each sample are the quadratic mean of these differences, as the Equation (1) shows.

$$\text{MAE} = \frac{\sum_{i=1}^{n} |y_i + x_i|}{n} = 0,021 \quad (2)$$

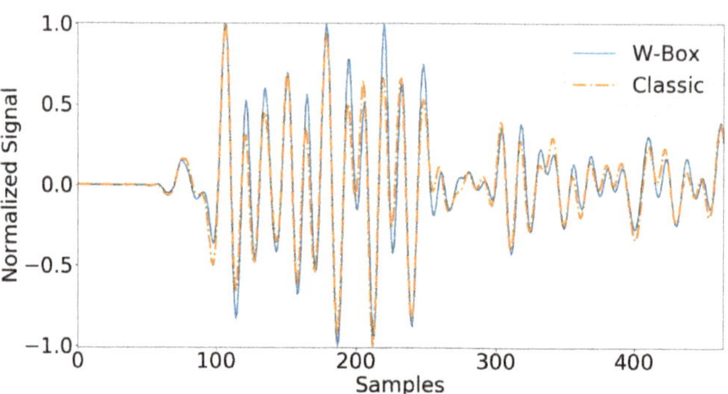

Figure 17. The comparison of normalized and preprocessed measurements recorded between the classic system and the W-Box.

Quite similar to RMSE, MAE (Equation (2)) is calculated with the average of the squares of the errors, corresponding to the expected value of the squared error loss. In both equations, x_i and y_i represent the value read on each of the compared measurement systems in the same time interval. n represents the total number of data measurements (for this measurement 5000 samples).

In order to assess the sensibility of the system, a white measurement was made to assess the noise floor of the system. Figure 18 shows the Fast Fourier Transform (FFT) of the raw signal in the frequency range of interest between 5 kHz to 110 kHz. The noise of the raw data is oscillating around ±1,22 mV, well below any actual signals.

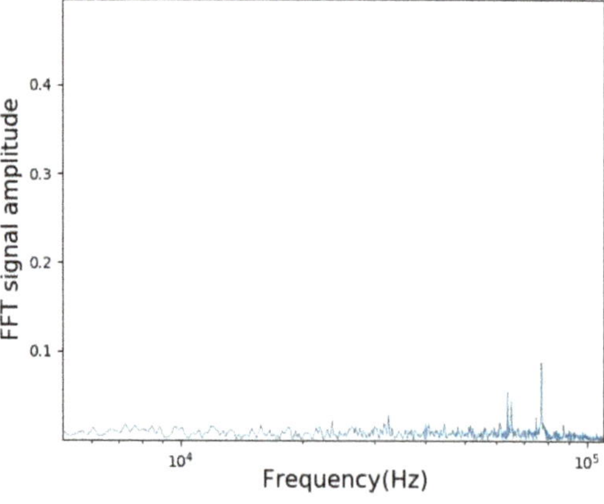

Figure 18. Noise floor of the W-Box. Fast Fourier Transform (FFT) of the white measurement in the frequency range of 5 kHz to 110 kHz.

According to [15,19,22], a signal to noise ratio of 2:1 is required to clearly identify the maximum of a signal. Therefore, we assume that the sensitivity of the W-Box system is twice the system noise. Considering the measured white noise raw data ±1.22 mV, the sensitivity is ±2.44 mV. After maximal amplification of 150-fold, a sensibility of approximately 16.2 µV is calculated.

To demonstrate the real measurement limits of the transducers with the W-Box, a preliminary field test was made. With full power and amplification, the signal to noise ratio was calculated and compared with the transducer's distance. This is shown in Figure 19.

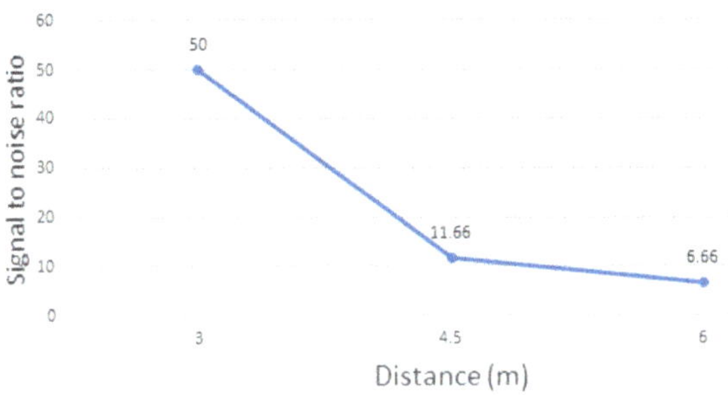

Figure 19. Signal to noise ratio vs. distance.

4. Discussion

The discussion focuses on the comparison between a reliable reference and the W-Box. The reliable reference used is a DAQmx from National Instruments and is in this paper referred as classic measurement device. The error results from the last section indicate 4.0% error with the RMSE method and 2.1% error with the MAE method. Regarding differences that spread the error, such as precision differences (14 bits for the W-Box ADC board and 16 bits for DAQmx) and different signal amplifiers, the calculated errors demonstrate satisfactory readability. The noise floor in Figure 18 is considerably low, despite the small peak at around 78 kHz.

The amplification limits in the field will be discussed in future articles with ongoing experiments. However, preliminary tests have shown that measurements at a distance of 6 meters can achieve a signal to noise ratio of 6.7. This value exceeds the maximum reached in previous experiments using the same sensors, such as [11,13,19].

One point to acknowledge is that the new system is comparatively slower than the traditional system. This is because it has multiple boards depending on the I2C protocol to communicate. The higher the demand for samples, the longer the time to perform a measurement, which can vary from 5 to 20 s per measurement pair. A possible solution to speed up the measurements is to change the communication protocol of the ADC board from I2C to SPI. Replacing the relays for solid state high-voltage transistors could also greatly increase the measurement speed. However, these are suggestions for the next generations of the W-Box if increasing the speed of measurement is necessary. Since the development of the temperature measurement system, several W-Boxes have been used in experiments. The temperature measurement system works well and with the calculated precision [23]. In a long-term bridge monitoring experiment, which was ongoing at the time this paper was written, the W-Box worked well even after several months. The system for sending data to the external server also worked correctly, as did the auto start system after a power outage in field tests.

Industrial equipment such as Consonic C2 (GS, ES and KS versions) from Geotron, V-Meter MK IV from James Instruments and Pundit Lab(+) from Prosec are available on the market. All of them exceed the W-box in acquisition speed and other technical areas. Nevertheless, the W-box has some significant advantages. Its design and the open source concept allow minimizing costs and simple reproduction, while the commercial systems are often more than 10,000 Euros, with limitations to raw data access, compatibility to different sensors and limited channels. The W-box can measure up to 75 channels, be easily adapted to ones needs and integrated in, e.g., student projects, and other academic and teaching areas. The compact and robust design, combined with the open linux operating system allows monitoring experiments in remote locations and easy and cost efficient maintenance. The W-Box is not considered an industrial equipment, as it has no industrial certifications and no protection against vibration or electromagnetic induction.

5. Conclusions

In terms of the objective of developing a system capable of ultrasonic monitoring with comparable performance to what is available on the market, the W-Box project has fulfilled the expectations. All project requirements were met and the durability and repeatability of the W-Box measurements, which have been used in research at BAM and our CoDA research partners for the past year, demonstrate system reliability. Just considering the parts and materials, the final cost for a 20-channel system with a temperature board is around € 450. The W-Box project will be published on GitHub with all the files for building the electronics and the necessary codes for the microprocessors and Raspberry Pi.

Author Contributions: Conceptualization, D.F.B.; methodology, D.F.B.; software, D.F.B.; validation, D.F.B. and N.E.; resources, E.N.; writing—original draft preparation, D.F.B.; writing—review and editing, N.E. and E.N.; visualization, D.F.B.; supervision, E.N.; project administration, E.N.; funding acquisition, E.N.; All authors have read and agreed to the published version of the manuscript.

Funding: This research was funded by the German Research Foundation (DFG) funding project FOR 2825.

Institutional Review Board Statement: not applicable.

Informed Consent Statement: not applicable.

Data Availability Statement: not applicable.

Acknowledgments: The W-Box concept is based on an idea of Herbert Wiggenhauser and was initiated as a student project of Florian Knopp, supervised by Frank Mielentz.

Conflicts of Interest: The authors declare no conflict of interest.

References

1. C597-02, A. Standard test method for pulse velocity through concrete. In *Annual Book of ASTM Standards, American Society for Testing and Materials*; ASTM International: West Conshohocken, PA, USA; 2002; Volume 4.
2. Reinhardt, H.W.; Grosse, C.U. *Report 31: Advanced Testing of Cement-Based Materials during Setting and Hardening-Report of RILEM Technical Committee 185-ATC*; RILEM Publications: Bagneux, France, 2005; Volume 31.
3. Thiele, M. Experimental Investigation and Analysis of Damage Evolution in Concrete Due to High-Cycle Fatigue Loading. Ph.D. Thesis, Technische Universität Berlin, Berlin, Germany, 2015.
4. Payan, C.; Garnier, V.; Moysan, J. Potential of nonlinear ultrasonic indicators for nondestructive testing of concrete. *Adv. Civil Eng.* **2010**, *2010*. doi:10.1155/2010/238472.
5. Planès, T.; Larose, E. A review of ultrasonic Coda Wave Interferometry in concrete. *Cem. Concr. Res.* **2013**, *53*, 248–255.
6. Legland, J.B.; Zhang, Y.; Abraham, O.; Durand, O.; Tournat, V. Evaluation of crack status in a meter-size concrete structure using the ultrasonic nonlinear coda wave interferometry. *J. Acoust. Soc. Am.* **2017**, *142*, 2233–2241.
7. Zhang, Y.; Planès, T.; Larose, E.; Obermann, A.; Rospars, C.; Moreau, G. Diffuse ultrasound monitoring of stress and damage development on a 15-ton concrete beam. *J. Acoust. Soc. Am.* **2016**, *139*, 1691–1701.
8. Hafiz, A.; Schumacher, T. Monitoring of stresses in concrete using ultrasonic coda wave comparison technique. *J. Nondestruct. Eval.* **2018**, *37*, 1–13.
9. Niederleithinger, E.; Wang, X.; Herbrand, M.; Müller, M. Processing ultrasonic data by coda wave interferometry to monitor load tests of concrete beams. *Sensors* **2018**, *18*, 1971.
10. Wang, X.; Niederleithinger, E.; Lange, M.; Stolpe, H. Implementation of Ultrasonic Coda Wave Interferometry on a Real Bridge. *Struct. Health Monitor.* **2019**, *2019*, doi:10.12783/shm2019/32365.
11. Niederleithinger, E.; Wolf, J.; Mielentz, F.; Wiggenhauser, H.; Pirskawetz, S. Embedded ultrasonic transducers for active and passive concrete monitoring. *Sensors* **2015**, *15*, 9756–9772. doi:10.3390/s150509756.

12. Knopp, F.; Mielentz, F.; Bernstein, T. Ultraschall-Messsystem für die Langzeitüberwachung von Betonkonstruktionen. In Proceedings of the DGZfP Jahrestagung 2019, Friedrichschafen, Germany, 27–29 May 2019.
13. Wang, X.; Chakraborty, J.; Bassil, A.; Niederleithinger, E. Detection of multiple cracks in four-point bending tests using the coda wave interferometry method. *Sensors* **2020**, *20*, 1986. doi:10.3390/s20071986.
14. The Robotics Back-End. Available online: https://roboticsbackend.com/raspberry-pi-3-pins/ (accessed on 15 January 2021).
15. Krautkrämer, J.; Krautkrämer, H. *Werkstoffprüfung mit Ultraschall*; Springer: Berlin, Germany, 2013.
16. Tietze, U.; Schenk, C. *Halbleiter-Schaltungstechnik*; Springer: Berlin, Germany, 2013.
17. Mielentz, F.; Krause, M.; Wüstenberg, H. Entwicklung einer Gruppenstrahler-Sendeeinheit für Ultraschalluntersuchungen von Betonbauteilen. *DGZfP-JAHRESTAGUNG 2002*. Available online: https://www.ndt.net/article/dgzfp02/papers/v44/v44.htm (accessed on 18 May 2021).
18. Epple, N.; Barroso, D.F.; Niederleithinger, E. Towards Monitoring of Concrete Structures with Embedded Ultrasound Sensors and Coda Waves—First Results of DFG for CoDA. In *European Workshop on Structural Health Monitoring*; Springer: Berlin, Germany, 2013; pp. 266–275.
19. Wolf, J. Schadenserkennung in Beton Durch Überwachung mit Eingebetteten Ultraschallprüfköpfen. Ph.D. Thesis, Universität Potsdam, Potsdam, Germany, 2017.
20. Hyndman, R.J.; Koehler, A.B. Another look at measures of forecast accuracy. *Int. J. Forecast.* **2006**, *22*, 679–688.
21. Pontius, R.G.; Thontteh, O.; Chen, H. Components of information for multiple resolution comparison between maps that share a real variable. *Environ. Ecol. Stat.* **2008**, *15*, 111–142. doi:10.1007/s10651-007-0043-y.
22. Merkblatt B 04. Ultraschallverfahren zur Zerstörungsfreien Prüfung im Bauwesen. *Deutsche Gesellschaft für Zerstörungsfreie Prüfung e.V. -DGZfP-, Fachausschuss Zerstörungsfreie Prüfung im Bauwesen, Unterausschuss Ultraschallprüfung*; 2018; ISBN 978-3-940283-98-6. Available online: https://www.baufachinformation.de/merkblatt-b-04/buecher/252875 (accessed on 18 May 2021).
23. Epple, N.; Fontoura Barroso, D.; Hau, J.; Epple, N. *Accounting for Long Term Environmental Influences on Ultrasonic Monitoring Measurements of Reinforced Concrete Constructions with Embedded Transducers*; Accepted for SHMII 10, July 2021, Porto, Portugal.

MDPI

Article

Design and Construction of A Novel Simple and Low-Cost Test Bench Point-Absorber Wave Energy Converter Emulator System

Ephraim Bonah Agyekum [1,*], Seepana PraveenKumar [1], Aleksei Eliseev [2] and Vladimir Ivanovich Velkin [1]

[1] Department of Nuclear and Renewable Energy, Ural Federal University Named after the First President of Russia Boris, 19 Mira Street, Ekaterinburg, 620002 Yeltsin, Russia; ambatipraveen859@gmail.com (S.P.); v.i.velkin@urfu.ru (V.I.V.)
[2] Ocean Rus Energy, Yubileynaya Street 1 Square 18, 623107 Pervouralsk, Russia; oceanrusenergy@gmail.com
* Correspondence: agyekumephraim@yahoo.com or agyekum@urfu.ru

Abstract: This paper proposed a test bench device to emulate or simulate the electrical impulses of a wave energy converter (WEC). The objective of the study is to reconstruct under laboratory conditions the dynamics of a WEC in the form of an emulator to assess the performance, which, in this case, is the output power. The designed emulator device is programmable, which makes it possible to create under laboratory conditions the operating mode of the wave generator, identical to how the wave generator would work under real sea conditions. Any control algorithm can be executed in the designed emulator. In order to test the performance of the constructed WEC emulator, an experiment was conducted to test its power output against that of a real point-absorber WEC. The results indicate that, although the power output for that of the real WEC was higher than the WEC emulator, the emulator performed perfectly well. The relatively low power output of the emulator was because of the type of algorithm that was written for the emulator, therefore increasing the speed of the motor in the algorithm (code) would result in higher output for the proposed WEC emulator.

Keywords: wave energy converter; emulator; point absorber; power take-off; hydrodynamics

Citation: Agyekum, E.B.; PraveenKumar, S.; Eliseev, A.; Velkin, V.I. Design and Construction of A Novel Simple and Low-Cost Test Bench Point-Absorber Wave Energy Converter Emulator System. *Inventions* **2021**, *6*, 20. https://doi.org/10.3390/inventions6010020

Received: 3 February 2021
Accepted: 18 March 2021
Published: 22 March 2021

1. Introduction

The need for a cleaner, cheaper and more reliable source of energy generation has led to increasing research in several renewable sources to assess their technoeconomic potential to meet global energy needs [1–6]. To meet this increasing demand for energy and also solve the issues related to reliability and availability, research on ocean wave energy has increased in recent times due to the huge energy potential it possesses [7,8]. Wave energy generation is at its initial stage of development and utilization; however, it presents lots of possibilities for the future due to its huge potential, relative to electricity production [9–11]. Research shows that the world's wave energy resources are about 17 TW h/year, with the highest average wave power located in the mid-latitudes, i.e., between 30° and 60° [9]. Offshore wave power levels averaged over the years vary from 30 to 100 kW/m and are located in latitudes of 40–50°, as well as smaller power levels further north and south. The average wave power in most tropical waters is below 20 kW/m [12].

Research toward the design and development of appropriate conversion devices has been progressive over the years. The design of a particular wave energy converter (WEC) is usually dependent on the location of its installation in the sea [13,14]. A survey of WECs showed a variety of designs, namely: attenuators, oscillating water columns, point absorbers, wave surge converters, overtopping devices and submerged point absorbers. Each of these systems has its own unique power take-off (PTO) mechanisms and can introduce fluctuations in grid voltage as a result of their variable power output when connected to a distribution grid. The variability associated with it is basically a function of the intensity of the wave for a given time and the dominant wave period [15]. There are, however, some barriers that have affected the development and use of marine energies, such as uncertainties associated with impacts from the marine environment on the wave farms [9,16,17], immature technology [9,18] and the fact that these technologies are regarded as economically unviable [19].

Experimental testing of WECs is key in its development; however, sea trials are too expensive and time consuming, and as a result, it ought to be preceded by small- to medium-scale testing. Even though this does not give a full representation of the performance of the prototypes accurately, it provides important information to developers, researchers and investors/entrepreneurs. Testing of models is a critical step in the development of offshore renewable energy technologies (RETs). It consists of challenges that demand guidance and experience. Hence, costly mistakes could result in a waste of resources and time [20].

A number of studies have either theoretically, experimentally or numerically assessed the performance of wave energy converters. Têtu [21] gave guidelines toward the development of WECs; their study also provided some guidance on model testing for WECs. Tongphong et al. [22] presented a novel WEC known as the ModuleRaft WEC comprising a floating modular flap and four rafts hinged at the main floating structure. They investigated the motion characteristics, optimization and performance of the model under regular wave conditions with the ANSYS AQWA model. Results from their study suggested that the ModuleRaft WEC operates best by utilizing all directions of wave energy by using a single-point mooring system. Al Shami et al. [23] studied the impact of increasing the total number of degrees of freedom by adding submerged objects to a point-absorber WEC. This was aimed at capturing more power with a lower resonant frequency with the same total mass and volume, similar to that of a piezoelectric vibration energy harvester.

Furthermore, Muthukumar and Jayashankar [24] employed a permanent magnet synchronous generator based on adjustable speed energy extraction to operate an air turbine near its maximum power point operation. The study considered an Indian wave energy plant, wave period, typical wave height and differential pressure/wave power variation to conduct an offline computation of direct current bus voltage dynamics using a MATLAB-based transient model of a complete WEC. Blanco et al. [25] proposed a laboratory test bench for testing a direct-drive linear generator for integration into a WEC. Their study focused on getting a model for an actuator to generate the behavior of a WEC in a real sea environment. Hazra et al. [26] presented a paper on emulating dynamic WEC behavior using a real-time simulator, as well as an induction motor under test bed conditions. The emulator was made up of a WEC using NI CompactRIO, an induction motor and an electric drive. Wahyudie et al. [27] designed a laboratory-scale WEC system grounded on a planar double-sided permanent magnet linear generator. Their aim was to produce an easy and simple method in the design and optimization of the various parameters to realize the intended output voltage with physical design constraints.

Testing of WEC devices in a real sea environment is usually designed using test data obtained from prototype WECs in laboratories. Nevertheless, very little information is usually available in relation to which electrical arrangement and control system is the optimum for a WEC, as well as the possible power output of the WEC in a particular location beforehand, although some simulations can be done. However, according to [28], such simulations fall short of real data, and this is the research gap this study seeks to bridge. The objective of the current study is to reconstruct under laboratory conditions the dynamics of a WEC in the form of an emulator to assess its performance; in this case, the output power and other parameters depending on the input algorithm. The proposed WEC emulator device is programmable, which makes it possible to create in laboratory conditions the operating mode of the wave generator, identical to how the wave generator would work under real sea conditions. Any control algorithm can be executed in the designed emulator. This is a unique construction, and it is expected to help in laboratory sections, reducing the time required for testing and also cost.

The rest of the paper is presented as follows: Section 2 covers a brief description of wave energy characteristics. The methodology and materials used for the study are presented in Section 3. Section 4 covers the results and discussion, and the conclusions are presented in Section 5.

2. Description of Wave Energy Characteristics

There are several mechanisms that lead to the generation of ocean waves, some of which include planetary forces or earthquakes; however, most of them are prompted by wind blowing on the surface of the fluid, known as wind waves. The ocean's depth and wavelength determine the velocity, which can be expressed using the dispersion relation (Equation (1)) [29]:

$$v = \sqrt{\frac{g\lambda}{2\pi}tanh\left(2\pi\frac{h}{\lambda}\right)} \tag{1}$$

where the wavelength is represented by λ, the water depth is denoted by h (m) and g (m/s^2) represents the acceleration due to gravity.

Research has shown that the diameter of the water particle motion circle declines with the water depth. Studies also indicate that nearly 95% of energy in waves can be found between the surface and the depth equivalent to a quarter of the deep-water wavelength [29,30]. A real sea wave may be regarded as a composition of several elementary waves with varying frequencies and directions, which is in contrast to that of a single-frequency sinusoidal wave moving in a specific direction. For a per unit area of the surface of a sea, stored energy equivalent to an average of Equation (2) is connected to the wave [12]:

$$E = \rho g H_{m0}^2/16 = \rho g \int_0^\infty S(f)df \tag{2}$$

where the mass density of the sea water is represented by ρ (kg/m^3), H_{m0} represents the significant wave height for the actual sea state and *S(f)* signifies the wave spectrum.

The total energy stored in waves is the summation of the potential E_p and kinetic E_k energies for every square meter, as expressed in Equation (3) [31].

$$E = E_p + E_k \tag{3}$$

The potential energy (PE) in relation to waves occurs as a result of the elevation of water, from the trough up to the crest. The average PE for each unit horizontal area can be computed using Equation (4) [31].

$$E_p = \frac{\rho g}{4}|A|^2 = \frac{\rho g}{16}H^2\left[\frac{J}{m^2}\right] \tag{4}$$

where A represents the amplitude, and H denotes the height.

The kinetic energy (KE) for each unit horizontal area can also be computed using Equation (5) [31].

$$E_k = \frac{\rho}{2}\omega^2|A|^2\int_{-\infty}^0 e^{2kz}dz = \frac{\rho}{2}\frac{\omega^2}{2k}|A|^2\left[\frac{J}{m^2}\right] \tag{5}$$

In situations where the deep-water conditions do not apply, $\omega^2 = gktanh(kh)$ is employed, hence the kinetic equation results in Equation (6) as

$$E_k = \frac{\rho}{4}|A|^2 gtanh(kh) = \frac{\rho}{16}H^2 gtanh(kh) \tag{6}$$

However, $\tanh(kh) \approx 1$ for deep water conditions, hence $\omega^2 = gk$, and the equation thus becomes Equation (7).

$$E_k = \frac{\rho g}{4}|A|^2 = \frac{\rho g}{16}H^2 \tag{7}$$

Therefore, the total energy in the waves results in Equation (8).

$$E = E_k + E_p = \frac{\rho g H^2}{16}(1 + tanh(kh)) \quad (8)$$

Equation (3) becomes Equation (9) under deep-water conditions.

$$E = E_k + E_p = \frac{\rho g H^2}{8} \quad (9)$$

2.1. Extraction of Wave Energy

The conversion of wave energy is a varying stochastic process as a result of radiation and diffraction; hence, the theory is usually device-based. There are various pneumatic and hydraulic power conversion devices that have been technologically advanced or proposed to transform the movements of waves into mechanical power [32]. WECs have rotating and reciprocating sections to use hydrodynamic lift forces generated by the flow over a lifting structure or hydrofoil, creating low speed output and high torque [32,33]. Every WEC device extracts some quantity of power from the waves according to its efficiency. Calculations for the power from waves are still not sufficiently explained as a result of the complexity and the stochastic progression of waves. The energy per wave period also known as the wave power density (W/m^2) can be calculated using Equation (10) [32].

$$P_{density} = \frac{pgA^2}{2T} \quad (10)$$

The power for each meter of the wave front and crest length can be expressed as in Equation (11) [32].

$$P_{wave\ front} = \frac{1}{8\pi}\rho g^2 A^2 T \quad (11)$$

$$P_{crest\ length} = \frac{1}{32\pi}\rho g^2 H^2 T \quad (12)$$

In the case of irregular waves, the approximate power per unit of wave front (kW/m) can be expressed as follows [32]

$$P_{wave\ front} \cong 0.42 H_s T_p \quad (13)$$

where T denotes the period (s), and the peak wave period (s) is represented by T_p. All other variables are as described supra.

Hydrodynamics of Point-Absorber WECs

Point-absorber WECs oscillate with either single or more degrees of freedom (DoF). Their linear dimensions are much smaller than usual wavelengths. The buoyancy and mass of point-absorber WECs are carefully chosen in order to resonate strongly with the waves. This type of WEC undertakes relative movements against a static reference [34]. As a general rule, in order to consider a WEC as a point absorber, its diameter has to be in the region of 5–10% of the prevailing wavelengths [35]. They can be grouped according to the DoF from which the ocean energy is captured. The performance of point absorbers is better as the frequency of the wave approaches its natural frequency. The most common WEC devices are the pitching and heaving bodies [34].

A heaving point-absorber WEC's behavior is comparable to that of a mechanical oscillator. It consists of a mass–spring–damper system with one DoF, which is exposed to an external force in the DoF's direction. [36]. Figure 1 shows a schematic of a point-absorber WEC. Two forces act on point-absorber WECs. These forces are both from the PTO and forces from external pressures on the buoy. This can be expressed as follows [31]:

$$ma = F_{pe} + F_{PTO} \quad (14)$$

where the acceleration is represented by α, the mass of the system is denoted by m, F_{pe} denotes the forces experienced by the WEC from the ocean and its waves as a result of external pressures and F_{PTO} represents the reaction forces emanating from the power take-off, which acts on the buoy.

$$F_{pe}(t) = F_e(t) + F_r(t) + F_{hs}(t) + F_d(t) \tag{15}$$

where the radiation force is represented by F_r, F_e denotes the excitation force. The drag force is also represented by F_d and F_{hs} is the hydrostatic or buoyancy force. The product of the total force and the velocity of the system yield the absorbed power of the oscillating system.

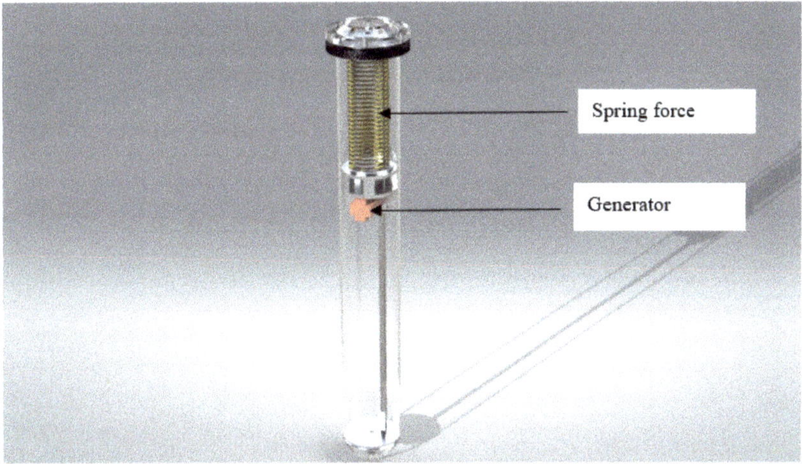

Figure 1. Schematic of a point-absorber WEC with direct mechanical PTO (designed by the authors).

The wave F_e exerted on a heaving point absorber is made up of both wave diffraction and Froude–Krylov forces. This force emanates from the incident wave striking the WEC's surface, which is held still in the water, arising from the potential flow wave theory. In order to calculate the wave F_e, it is required to integrate both the diffracted wave potential and the incident wave pressure (i.e., Froude–Krylov) over the surface of the WEC, as indicated in Equation (16) [37].

$$\overrightarrow{F_{wave\ excitation}} = \iint_{wetted\ surface}^{body} p_{wave}\hat{n}dS \tag{16}$$

where the pressure of both the diffracted and incident wave potentials are denoted by p_{wave}, $\hat{n}$ represents the unit vector and the wetted surface of the oscillating bodies is denoted by S.

It is also possible to use the finite-element method to solve the integral around the limits of the oscillating systems; on the other hand, in the linear domain, it is assumed that the wave F_e is the oscillatory force proportional to the elevation of the incoming wave. This can be represented as follows [37]:

$$F_e = AF_{ex(\omega)}e^{i\omega t+\phi_\omega} \tag{17}$$

In this equation, A is the wave amplitude, the complex amplitude of the Froude–Krylov and diffraction wave excitation forces is represented by $F_{ex(\omega)}$, ω is the wave angular frequency in rad/s, the phase angle between the excitation force and the incoming wave is denoted by ϕ_ω.

With respect to the time domain, one can model the wave F_e of a first-order regular wave using Equation (18).

$$F_{(t)}^{wave} = F_{ex(\omega)} A cos\left(wt + \varnothing_{(w\omega)}\right) \quad (18)$$

Under the time domain, one is able to simulate wave forces under the excitation of irregular waves, which can be achieved through the superposition of N different sinusoidal irregular waves, i.e., starting from $n - 1$ to n = N. This is modeled as follows:

$$F_{(t)}^{wave} = \sum_{n=1}^{N} F_{ex(\omega_n)} A_n cos\left(w_n t + \varnothing_{(\omega_n)} + \varphi_n\right) \quad (19)$$

where φ_n represents an arbitrarily selected phase value for the wave elevation between [0, 2π], and A_n denotes the calculated wave amplitude via the mean square value from an irregular wave spectrum, e.g., the JONSWAP spectrum [37].

Radiation force is evaluated on the assumption that the incident wave set is zero. The water surface is still, and the WEC (point absorber) oscillates on its surface, generating radiated waves, which react with the point absorber as radiation forces. The forces are computed using Equation (20) [37,38].

$$\overrightarrow{F_{radiation}} = \iint_{wetted\ surface}^{body} p_{radiated\ wave} \hat{n} dS \quad (20)$$

For a frequency-dependent domain, the radiation force is [37,38]:

$$F_{radiation} = c_{r(\omega)} \dot{y} + m_{a(\omega)} \ddot{y} \quad (21)$$

The noncausality of radiation damping forces was presented by Falnes [39] for the time domain, as indicated in Equation (22).

$$F_{radiation} = m_a^{\infty} \ddot{y}_{(t)} + \int_{-\infty}^{t} RIF(t-\tau) \dot{y}_{(t)} d\tau \quad (22)$$

where $RIF_{(t)}$ represents the radiation impulse function in time domain, $c_{r(\omega)}$ denotes the radiation damping coefficient and $m_{a(\omega)}$ represents the added mass.

The equation solution, differential equations and hydrodynamics can be solved by several methods in the frequency domain and in a semidiscretization way. Urbikain et al. [40] proposed interesting approaches in that line by giving steps involving solving such equations or the group stability prediction in the straight turning of a flexible workpiece by the collocation method.

3. Materials and Methodology

The methodology and description of the operations of the proposed WEC emulator are presented in this section. The experimental approach adopted to validate the designed emulator is also provided in this section.

3.1. Emulation Technique

The device includes stepper motors because wave generators extract electrical energy from sea waves and potential and kinetic energy contained in the structure of sea waves. Wave generators convert it into mechanical energy, then mechanical energy is converted into electrical energy due to the presence of stepper motors on the board of the wave generators. The emulator was designed and constructed using the various components in Table 1. The schematic diagram for the emulator is illustrated in Figure 2.

Table 1. Components of the designed emulator.

Number	Designation	Quantity
1	Lead stepper motor	1
2	Driven stepper motor	1
3	RS-25 power supply	1
4	Power supply LRS-350-48	1
5	G210 Driver	1
6	Arduino Nano	1
7	Network socket C14 on REX ANT housing	1
8	Rocker switch 250 V 16 A	1
9	Panel socket 10-0019 red	2
10	Panel socket 10-0019 black	2
11	Switch-push button 250 V 1 A yellow	1
12	Switch-push button 250 V 1 A green	1
13	Switch-push button 250 V 1 A blue	1
14	**Liquid-crystal display (LCD)**	1

The leading stepper motor in the test bench is designed to simulate the trajectory, amplitude, period and other parameters of sea waves. The initial data for programming the operation of the lead stepper motor in the test bench are taken from a reference wave generator operating in real marine conditions. The initial indicators that are taken from the operation of the wave generator are time, sea wave height and electrical impulse. After that, with the initial data available, a mathematical calculation is performed.

The installed stepper motor in the wave generator on the shaft has a gear with a diameter of 32 mm. Since the wave generator is equipped with a torsion (spring) pendulum, the reciprocating movements of the stepper motor inside the wave generator come into resonance with the sea wave. Knowing the period of time and the height of the movement of the wave generator in space under sea conditions, as well as knowing the diameter of the gear on the shaft of the stepper motor, we calculated the number of revolutions R_n made by the stepper motor using Equation (23). We thus obtained the length of one revolution performed by a stepper motor. Further, the height of the movement in space by the wave generator is divided by the length of one revolution, after which we obtained the number of revolutions made by the stepper motor within the wave revolution.

$$R_n = 32 \cdot \pi \tag{23}$$

In the emulator, the initial data obtained are entered by writing a program for the Arduino board. The program is written on the lead stepper motor. When the device is started in the specified program, it allows obtaining a 100% reproduction of the operations of the stepper motor included in the wave generator experienced under real sea conditions on the slave stepper motor. The proposed emulator is indicated in Figure 3.

3.2. Experimental Setup

The experimental process is described in this section to assess the effectiveness and performance of the built emulator. The testing of the performance of the designed and constructed WEC emulator was achieved with the help of an APPA 109 N multimeter, as well as a laptop to record the values of the experiment, as shown in Figure 4. The output from the experimental work is presented in the results and discussions section.

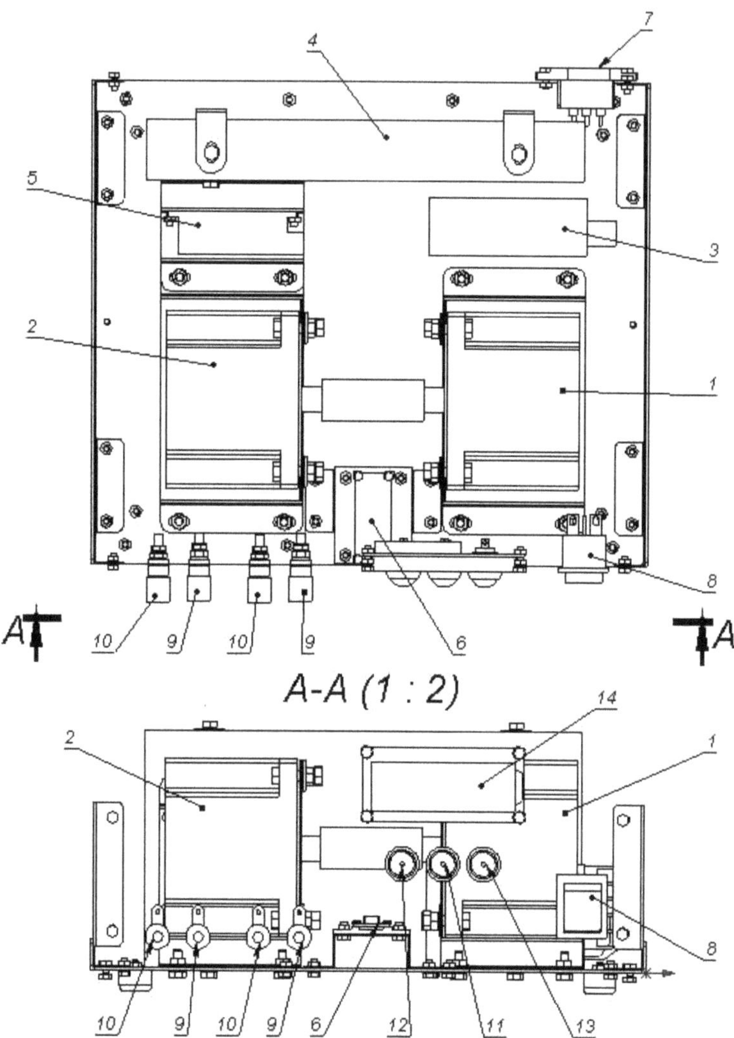

Figure 2. Schematic diagram of the WEC emulator.

3.3. *Logic of the Motor Rotation*

Figure 5 shows the motor rotation logic. It was created as input for programming the master stepper motor. In the first column, the total cycle (period) is in 8 s. The loop is then closed, repeating every second equally. The cycle of seconds is divided into parts, and each part is equal to 1 s. Every 4 s in a loop is absolute 0. The cycle begins: 1 s = 1.5 revolutions per second in the left direction (positive), followed by 2 s = 2.5 revolutions made per second to the left (positive), 3 s = 1.5 revolutions per second to the left (positive), 4 s = 0 of the completed revolution (stop), 5 s = 1.5 revolutions per second in the direction to the right side (negative), 6 s = 2.5 revolutions per second towards the right side (negative), 7 s = 1.5 revolutions per second in the direction to the right (negative) and 8 s is equal to the complete turnover in 1 s (stop).

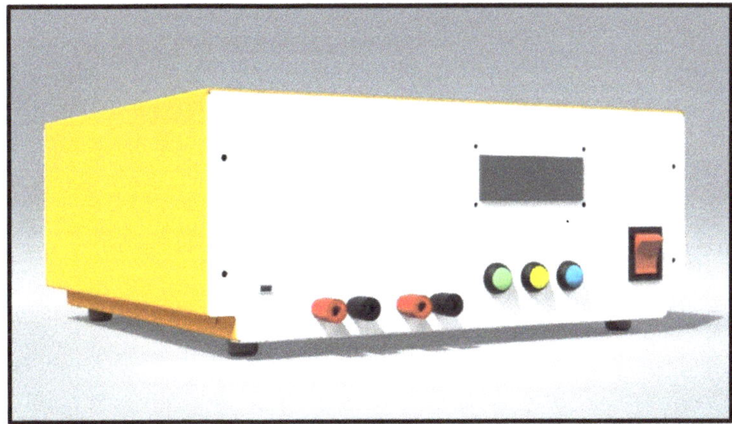

Figure 3. Proposed WEC emulator to be constructed.

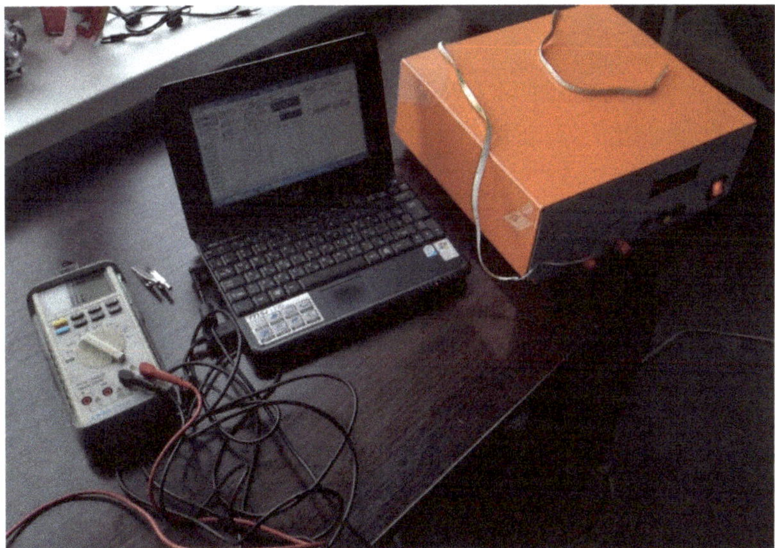

Figure 4. Experimental setup made up of a (from left to right) multimeter, laptop computer and the newly built WEC emulator.

3.4. Experimental Setup for the Real Point-Absorber WEC

In order to validate the results from the newly built WEC emulator, another experiment using a point-absorber WEC constructed by one of the authors located in the Ural Federal University laboratory was performed. Wave generation for the WEC was performed using a special paddle developed by the university, which is capable of generating wave motions with a frequency ranging between 0.5 and 2 Hz. The experimental setup for the WEC is represented in Figure 6. Data were recorded with the help of the APPA software. The WEC used in this study uses the direct mechanical drive PTO system. The direct mechanical drive PTO system translates the mechanical energy of the oscillating body as a result of the movement of the waves into electricity using an extra mechanical system, which drives a rotary electrical generator. One of the advantages associated with this PTO technology is that it has high efficiency because only up to three energy conversions are necessary [21]. The experiment considered only the heave motion of the wave.

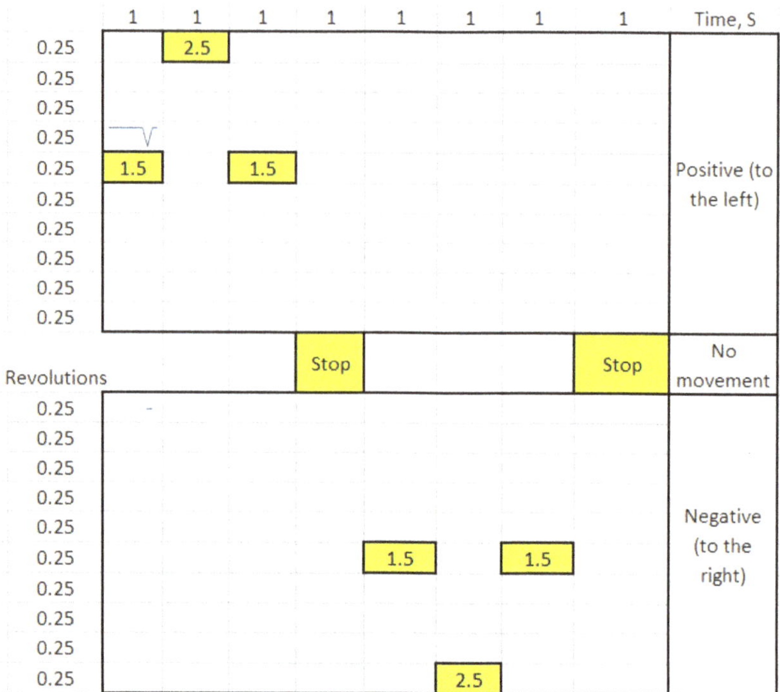

Figure 5. Motor rotation logic.

Figure 6. Experimental setup for the point-absorber WEC generator.

4. Results and Discussion

The results and discussion from the experiments are presented in this section. The emulator device is programmable, which makes it possible to create under laboratory conditions the operating mode of the wave generator, identical to how the wave generator would work in real sea conditions. It allows the user to set different programs, with different amplitudes, heights, periods and other initial values of sea waves as a renewable energy source. The test bench, according to a given program, will generate electrical impulses that are close to real conditions as much as possible. This allows in laboratory conditions to conduct scientific, educational, engineering and design work on the processing of electrical impulses and construction (design) of a further electronic circuit of wave generators.

In assessing the performance of WEC, one of the parameters of major concern is its average output power because the total energy production is associated with the average power rather than the maximum power. The average absorbed power over a single wave period for a sinusoidal wave is equivalent to half of the maximum power [41]. The performance of the emulator relative to its voltage output is presented in Figure 7b, which is compared to the real WEC output in Figure 7a. Figure 8a,b also indicates the power output of the emulator and WEC, respectively. The results from the two figures show a significant disparity between the WEC and the emulator. The power output for the WEC is higher than that of the emulator. This disparity is a result of the differences in the speed level of the motor. The program (code) written to operate the emulator has a slower speed with respect to the operations of the motor. As a result, the power output of the emulator is lower. This is not because the emulator cannot perform, but it is a result of the program written for the WEC emulator, and increasing the speed of the motor will positively affect its output power. The power output of the generator in the speed control mode can be employed to evaluate the power conversion circuit. Nevertheless, the generator operates as an infinite power source. Hence, in order to assess the WEC's generation capability and also confirm the control operation of the generator, it is necessary to emulate the WEC through the torque control of the motor [26].

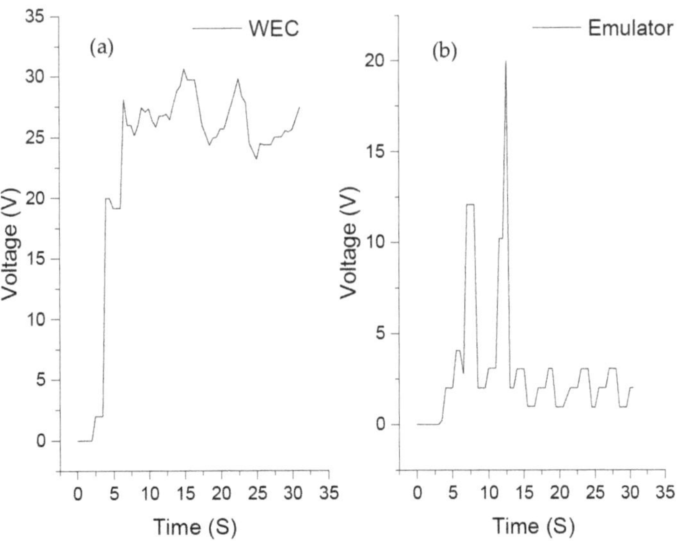

Figure 7. Output voltage for (**a**) WEC and (**b**) emulator.

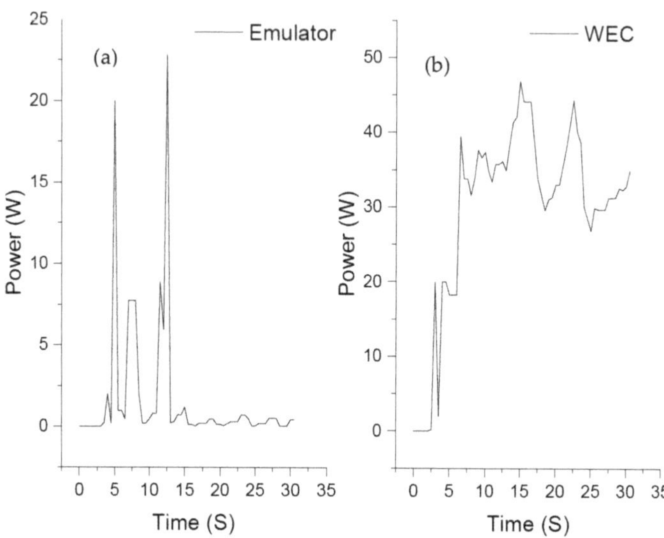

Figure 8. Output power for (**a**) emulator (**b**) WEC.

As indicated in Figure 8, both systems provided fluctuated power, which reflects the behavior of an incident wave. In the case of a direct-drive WEC, the fluctuations can be smoothened with the integration of an energy storage system. The appropriate technology, which seems to be the best relative to the flattening of the power of such systems, is the inclusion of supercapacitors [42], which future studies by authors are going to consider.

The results from the experimental analysis are confirmed by an earlier study with the same WEC [43], which identified that the power capacity of the unit is within the range of 20 to 60 W. According to the study, the power of a WEC can be increased when a number of WECs are combined to form a single cluster.

5. Conclusions

A WEC emulator was developed and presented in this paper. This emulator was constructed to serve as a substitute for a WEC in laboratories for the performance assessment of a WEC as if it were in a real sea environment. This is expected to help researchers in the wave energy industry to cut down the cost and time associated with the testing of WECs in a real sea environment. The main objective of designing and constructing such a system has been achieved. An experiment was conducted with a WEC emulator based on an initial program (algorithm) that was written to test its performance. The outcome was compared with experimental data from a mini point-absorber WEC designed and patented by one of the authors of this paper. The results showed that the newly built WEC emulator has a relatively small output power compared to a real WEC. This can be attributed to the speed of the motor in the program written for the emulator. Increasing the speed will increase the output power of the emulator. This means that the proposed WEC emulator is an important tool that can be used in the wave energy industry to assess the effectiveness of a designed WEC prior to construction and deployment. The emulator has been designed in such a way that any control algorithm can be executed in it.

This work is, however, the initial aspect of the development and assessment of the current project. There are other future studies that have to be conducted to test the performance of the emulator further. Such studies will include the simulation of the developed emulator in some engineering tools such as MATLAB to be able to conduct a more dynamic assessment of it. This will help to obtain more results to compare with other models. However, the main objective of building the model has been achieved. With respect to the device, the following are the future plans:

1. At the moment, work is underway to modernize the test device. A number of changes have been made to the design documentation aimed at a significant reduction in the size of the device. This set of measures should make it possible to reduce the cost of manufacturing the device during serial production.

2. It is planned to create one more additional program of the device operation algorithm.

3. The next step is to create a menu program for the device interface, which will allow the first and second programs of the operating mode code, as well as several more programs, to be simultaneously placed in the device.

6. Patents

The prototype WEC in this paper, as indicated in Figure 1, used for the experimental analysis is protected under Russian Law (Patent Number: 139434).

Author Contributions: Conceptualization, A.E.; methodology, A.E. and E.B.A.; software, A.E., E.B.A. and S.P.; validation, A.E. and E.B.A.; formal analysis, A.E. and E.B.A.; investigation, A.E. and E.B.A.; resources, A.E. and E.B.A., S.P.; data curation, A.E., E.B.A. and S.P.; writing—original draft preparation, E.B.A.; writing—review and editing, E.B.A.; visualization, E.B.A.; supervision, V.I.V.; project administration, A.E. and E.B.A. All authors have read and agreed to the published version of the manuscript.

Funding: This research received no external funding.

Institutional Review Board Statement: Not applicable.

Informed Consent Statement: Not applicable.

Data Availability Statement: The data presented in this study are available on request from the corresponding author.

Conflicts of Interest: The authors declare no conflict of interest.

References

1. Agyekum, E.B. Energy poverty in energy rich Ghana: A SWOT analytical approach for the development of Ghana's renewable energy. *Sustain. Energy Technol. Assess.* **2020**, *40*, 100760. [CrossRef]
2. Agyekum, E.B. Techno-economic comparative analysis of solar photovoltaic power systems with and without storage systems in three different climatic regions, Ghana. *Sustain. Energy Technol. Assess.* **2021**, *43*, 100906. [CrossRef]
3. Faraggiana, E.; Chapman, J.; Williams, A.; Masters, I. Genetic based optimisation of the design parameters for an array-on-device orbital motion wave energy converter. *Ocean Eng.* **2020**, *218*, 108251. [CrossRef]
4. Agyekum, E.B.; Velkin, V.I. Optimization and techno-economic assessment of concentrated solar power (CSP) in South-Western Africa: A case study on Ghana. *Sustain. Energy Technol. Assess.* **2020**, *40*, 100763. [CrossRef]
5. Agyekum, E.B.; Nutakor, C. Feasibility study and economic analysis of stand-alone hybrid energy system for southern Ghana. *Sustain. Energy Technol. Assess.* **2020**, *39*, 100695. [CrossRef]
6. Agyekum, E.; Velkin, V.I.; Hossain, I. Sustainable energy: Is it nuclear or solar for African Countries? Case study on Ghana. *Sustain. Energy Technol. Assess.* **2020**, *37*. [CrossRef]
7. Isaacs, J.D.; Castel, D.; Wick, G.L. Utilization of the energy in ocean waves. *Ocean Eng.* **1976**, *3*, 175–187. [CrossRef]
8. Li, X.; Chen, C.; Li, Q.; Xu, L.; Liang, C.; Ngo, K.; Parker, R.G.; Zuo, L. A compact mechanical power take-off for wave energy converters: Design, analysis, and test verification. *Appl. Energy* **2020**, *278*, 115459. [CrossRef]
9. Astariz, S.; Iglesias, G. The economics of wave energy: A review. *Renew. Sustain. Energy Rev.* **2015**, *45*, 397–408. [CrossRef]
10. Akpınar, A.; Kömürcü, M.I. Assessment of wave energy resource of the Black Sea based on 15-year numerical hindcast data. *Appl. Energy* **2013**, *101*, 502–512. [CrossRef]
11. Iglesias, G.; Carballo, R. Choosing the site for the first wave farm in a region: A case study in the Galician Southwest (Spain). *Energy* **2011**, *36*, 5525–5531. [CrossRef]
12. Falnes, J. A review of wave-energy extraction. *Mar. Struct.* **2007**, *20*, 185–201. [CrossRef]
13. Hazra, S.; Bhattacharya, S. Modeling and Emulation of a Rotating Paddle Type Wave Energy Converter. *IEEE Trans. Energy Convers.* **2017**, *33*, 594–604. [CrossRef]

14. Hazra, S.; Shrivastav, A.S.; Gujarati, A.; Bhattacharya, S.; Hazra, S.; Gujarati, A.; Bhattacharya, S. Dynamic emulation of oscillating wave energy converter. In Proceedings of the 2014 IEEE Energy Conversion Congress and Exposition (ECCE), Pittsburgh, PA, USA, 14–18 September 2014; pp. 1860–1865.
15. Biligiri, K.; Harpool, S.; Von Jouanne, A.; Amon, E.; Brekken, T. Grid emulator for compliance testing of Wave Energy Converters. In Proceedings of the 2014 IEEE Conference on Technologies for Sustainability (SusTech), Portland, OR, USA, 24–26 July 2014; pp. 30–34.
16. Abanades, J.; Greaves, D.M.; Iglesias, G. Wave farm impact on the beach profile: A case study. *Coast. Eng.* **2014**, *86*, 36–44. [CrossRef]
17. Frid, C.; Andonegi, E.; Depestele, J.; Judd, A.; Rihan, D.; Rogers, S.I.; Kenchington, E. The environmental interactions of tidal and wave energy generation devices. *Environ. Impact Assess. Rev.* **2012**, *32*, 133–139. [CrossRef]
18. López, I.; Iglesias, G. Efficiency of OWC wave energy converters: A virtual laboratory. *Appl. Ocean Res.* **2014**, *44*, 63–70. [CrossRef]
19. Leijon, M.; Danielsson, O.; Eriksson, M.; Thorburn, K.; Bernhoff, H.; Isberg, J.; Sundberg, J.; Ivanova, I.; Sjöstedt, E.; Ågren, O.; et al. An electrical approach to wave energy conversion. *Renew. Energy* **2006**, *31*, 1309–1319. [CrossRef]
20. Portillo, J.; Collins, K.; Gomes, R.; Henriques, J.; Gato, L.; Howey, B.; Hann, M.; Greaves, D.; Falcão, A. Wave energy converter physical model design and testing: The case of floating oscillating-water-columns. *Appl. Energy* **2020**, *278*, 115638. [CrossRef]
21. Têtu, A. Power Take-Off Systems for WECs. In *Quantitative Monitoring of the Underwater Environment*; Springer International Publishing: Cham, Switzerland, 2016; pp. 203–220. [CrossRef]
22. Tongphong, W.; Kim, B.-H.; Kim, I.-C.; Lee, Y.-H. A study on the design and performance of ModuleRaft wave energy converter. *Renew. Energy* **2021**, *163*, 649–673. [CrossRef]
23. Al Shami, E.; Wang, X.; Ji, X. A study of the effects of increasing the degrees of freedom of a point-absorber wave energy converter on its harvesting performance. *Mech. Syst. Signal Process.* **2019**, *133*. [CrossRef]
24. Muthukumar, S.; Jayashankar, V. Micro-controller based emulation of a wave energy converter. In Proceedings of the 2011 International Conference on Emerging Trends in Electrical and Computer Technology, Nagercoil, India, 23–24 March 2011; pp. 186–189.
25. Blanco, M.; Moreno-Torres, P.; Lafoz, M.; Beloqui, M.; Castiella, A. Development of a laboratory test bench for the emulation of wave energy converters. In Proceedings of the 2015 IEEE International Conference on Industrial Technology (ICIT), Seville, Spain, 17–19 March 2015; pp. 2487–2492.
26. Hazra, S.; Kamat, P.; Shrivastav, A.; Bhattacharya, S. Emulation of oscillating wave energy converter for laboratory test bed. In Proceedings of the 3rd Marine Energy Technology Symposium, Washington, DC, USA, 27–29 April 2015; p. 8.
27. Wahyudie, A.; Susilo, T.B.; Jama, M.; Mon, B.F.; Shaaref, H. Design of a double-sided permanent magnet linear generator for laboratory scale ocean wave energy converter. In Proceedings of the OCEANS 2017—Anchorage, Anchorage, AK, USA, 18–21 September 2017; pp. 1–5.
28. Duquette, J.; O'Sullivan, D.; Ceballos, S.; Alcorn, R. Design and Construction of an Experimental Wave Energy Device Emulator Test Rig. In Proceedings of the European Wave and Tidal Energy Conference, Uppsala, Sweden, 7 September 2009.
29. Wang, L.; Isberg, J.; Tedeschi, E. Review of control strategies for wave energy conversion systems and their validation: The wave-to-wire approach. *Renew. Sustain. Energy Rev.* **2018**, *81*, 366–379. [CrossRef]
30. Ilyas, A.; Kashif, S.A.; Saqib, M.A.; Asad, M.M. Wave electrical energy systems: Implementation, challenges and environmental issues. *Renew. Sustain. Energy Rev.* **2014**, *40*, 260–268. [CrossRef]
31. Steinn, G. *Hydrodynamic Investigation of Wave Power Buoys*; KTH Royal Institute of Technology: Stockholm, Sweden, 2013.
32. Yuce, M.I.; Muratoglu, A. Hydrokinetic energy conversion systems: A technology status review. *Renew. Sustain. Energy Rev.* **2015**, *43*, 72–82. [CrossRef]
33. Rourke, F.O.; Boyle, F.; Reynolds, A. Marine current energy devices: Current status and possible future applications in Ireland. *Renew. Sustain. Energy Rev.* **2010**, *14*, 1026–1036. [CrossRef]
34. Faizal, M.; Ahmed, M.R.; Lee, Y.-H. A Design Outline for Floating Point Absorber Wave Energy Converters. *Adv. Mech. Eng.* **2014**, *6*. [CrossRef]
35. Falnes, J.; Lillebekken, P.M. Budal's latching-controlled-buoy type wave-power plant. In Proceedings of the 5th European Wave Energy Conference; 2003; p. 12. Available online: https://core.ac.uk/download/pdf/30849835.pdf (accessed on 18 March 2021).
36. Backer, G.D. *Hydrodynamic Design Optimization of Wave Energy Converters Consisting of Heaving Point Absorbers*; Ghent University: Ghent, Belgium, 2009.
37. Al Shami, E.; Zhang, R.; Wang, X. Point Absorber Wave Energy Harvesters: A Review of Recent Developments. *Energies* **2018**, *12*, 47. [CrossRef]
38. Sjökvist, L.; Göteman, M.; Rahm, M.; Waters, R.; Svensson, O.; Strömstedt, E.; Leijon, M. Calculating buoy response for a wave energy converter—A comparison of two computational methods and experimental results. *Theor. Appl. Mech. Lett.* **2017**, *7*, 164–168. [CrossRef]
39. Falnes, J. On non-causal impulse response functions related to propagating water waves. *Appl. Ocean Res.* **1995**, *17*, 379–389. [CrossRef]
40. Urbikain, G.; De Lacalle, L.L.; Campa, F.; Fernández, A.; Elías, A. Stability prediction in straight turning of a flexible workpiece by collocation method. *Int. J. Mach. Tools Manuf.* **2012**, *54-55*, 73–81. [CrossRef]

41. Hai, L.; Svensson, O.; Isberg, J.; Leijon, M. Modelling a point absorbing wave energy converter by the equivalent electric circuit theory: A feasibility study. *J. Appl. Phys.* **2015**, *117*, 164901. [CrossRef]
42. Aubry, J.; Ben Ahmed, H.; Multon, B.; Babarit, A.; Clément, A. Wave Energy Converters. In *Marine Renewable Energy Handbook*; John Wiley & Sons, Inc.: Hoboken, NJ, USA, 2013; pp. 323–366.
43. Hossain, I.; Velkin, V.I.; Shcheklein, S.E.; Eliseev, A.V. Structural Design Development of a Float Type Wave Micro Power Plant. *IOP Conf. Ser. Mater. Sci. Eng.* **2019**, *481*, 012007. [CrossRef]

MDPI
St. Alban-Anlage 66
4052 Basel
Switzerland
Tel. +41 61 683 77 34
Fax +41 61 302 89 18
www.mdpi.com

Inventions Editorial Office
E-mail: inventions@mdpi.com
www.mdpi.com/journal/inventions

www.ingramcontent.com/pod-product-compliance
Lightning Source LLC
LaVergne TN
LVHW070706170726
843515LV00004B/504

* 9 7 8 3 0 3 6 5 2 9 8 9 9 *